MW00814934

NEW DEVELOPMENTS IN OPTICS RESEARCH

LASERS AND ELECTRO-OPTICS RESEARCH AND TECHNOLOGY

Additional books in this series can be found on Nova's website under the Series tab.

Additional E-books in this series can be found on Nova's website under the E-book tab.

LASERS AND ELECTRO-OPTICS RESEARCH AND TECHNOLOGY

NEW DEVELOPMENTS IN OPTICS RESEARCH

MATTHEW P. GERMANNO
EDITOR

Nova Science Publishers, Inc.
New York

Library of Congress Cataloging-in-Publication Data

New developments in optics research / editor, Matthew P. Germanno.
    p. cm.
  Includes index.
  ISBN 978-1-61324-505-7 (hardcover)
  1. Optics--Research. I. Germanno, Matthew P.
  QC363.N49 2011
  535--dc23
                        2011012977

Published by Nova Science Publishers, Inc. † New York

CONTENTS

PREFACE

This book presents current research in the study of optics, with a particular focus on sensors and fibers. Topics discussed include optical current sensors; evanescent field tapered fiber optic biosensors; new challenges in raman amplification for fiber communication systems; fiber bragg gratings in high birefringence optical fibers and optics research applied to the turin shroud.

Chapter 1- In comparison with electric current transformers, an optical current sensor or transformer (OCT) has the invaluable features of non-contact, highly insulating and high-speed measurements. Because the OCT technology is closely related to the optical magnetic-field sensing technology, a lot of optical magnetic-field sensors are potentially applicable to OCT as well. The most reliable OCT, however, should be based on Ampere's circuital law, according to which the line integration of current-induced magnetic fields along a closed loop is equal to the current which intersects the closed loop. In order to perform this integration optically, the authors use Faraday effects, which states that the plane of linear polarization of a light beam propagating in an isotropic material subject to magnetization is rotated in proportion to the magnetization. So far many works on OCT have been made from engineering as well as scientific interests, and some OCTs developed have reached the stage of practical use and are commercially available. This chapter is devoted to the description of the present state-of-the-art of the OCT technology. The basic principles, types, performances and applications of OCTs are addressed.

Chapter 2- Tapered fiber optic biosensors (TFOBS) are made from optical fibers, and, are capable of detecting specific analytes using optical responses. They have been used for the measurement of physical and chemical properties of biological molecules and have several applications including environmental monitoring, drug screening, clinical diagnostics, and defense. TFOBS offer many advantages including flexibility, ease of use, affordability, and ability to perform sensing using a small amount of sample. These sensors are based on the evanescent field associated with fiber, and, often are also referred to as Evanescent Field Tapered Fiber Sensors. In this chapter, the basics of TFOBS are discussed, along with an up-to-date literature review of TFOBS. Experimental methods and recent results from the authors' laboratory are also presented.

Chapter 3- The deployment of optical communication systems through long haul networks required the development of transparent optical amplifiers, for replacement of the expensive and limitative optoelectronic regeneration. The increasing distance between amplification sites saves amplification huts reducing by this way the investment and operational cost in the network management.

The first choice for transparent optical amplification pointed out to the Erbium Doped Fiber amplifiers (EDFA), which was a mature technology by the beginning of the last decade of the XX century. However, the growing demand in terms of transmission capacity has been increasing dramatically, fulfilling the entire spectral band of the EDFA, and wideband amplifiers are now required. Raman fiber amplifiers (RFA) have emerged as a key technology for the optical networks.

Chapter 4- The development of the fiber optical technology was an important step in the revolution of global communications and in information technology. One of these developments happened in the 70's with the first optical fibers with low attenuation, a feature that enabled long- distance communication with high bandwidth. The intrinsic optical bandwidth of the optical fibers has also allowed the propagation of different simultaneous channels, allowing the transmission of data at Tbit/s rates. In these systems, in addition to transmission and amplification, it is often necessary to do all-optical processing to the signal. This is due to the inherent advantages of the optical processing, relative to the optic-electric-optic processing, like the higher flexibility to operate at different bit rates and modulation formats and also at the higher bandwidth. The evolution of the fiber optical technology has also enabled the development of devices for all optical processing. In this way, the insertion loss is reduced and the processing quality improved. One of the factors contributing to all-fiber optical processing devices was the discovery of the photosensitivity in optical fibers. It was documented for the first time in 1978 and led to the development of fiber Bragg gratings (FBG).

Chapter 5- The word "shroud" corresponds to the Italian "Sindone" that derives from the ancient Greek and it means "burial garment in which a corpse is wrapped". The Turin Shroud (TS) is 4.4 m long and 1.1 m wide, on which the complete front and dorsal images of the body of a man are indelibly impressed.

The TS is believed by many to be the burial cloth in which Christ was wrapped before being placed in a tomb in Palestine about 2000 years ago and the Science has not demonstrated the contrary. It is the most important Relic of Christianity and, of all religious relics, it has generated the greatest controversy. From a scientific point of view, this Relic is still unexplainable because up to now the body image has not been reproduced in all its details.

In: New Developments in Optics Research
Editor: Matthew P. Germanno

ISBN: 978-1-60324-505-7
© 2012 Nova Science Publishers, Inc

Chapter 1

OPTICAL CURRENT SENSORS

Toshihiko Yoshino[1] and Masayuki Yokota[2]

[1]The Kaisei Academy, Nishi-Nippori 4-2-4, Arakawa-ku,
Tokyo 116-0013, Japan
[2]Department of Electronic Engineering, Faculty of Engineering,
Gunma University, 1-5-1 Tenjin-cho, Kiryu,
Gunma 376-8515, Japan

ABSTRACT

This review article describes the present state-of-the-art of fiber optic current sensing technology based on Faraday effect. The importance of the fulfillment of HICOC (homogeneous isotropic closed optical circuit) condition is addressed. The optical current sensors are categorized into three types, viz. all-fiber, bulk-optic and hybrid, and their available HICOC techniques, advantageous features, performances and applications are mentioned.

1. INTRODUCTION

In comparison with electric current transformers, an optical current sensor or transformer (OCT) has the invaluable features of non-contact, highly insulating and high-speed measurements [1,2]. Because the OCT technology is closely related to the optical magnetic-field sensing technology, a lot of optical magnetic-field sensors are potentially applicable to OCT as well. The most reliable OCT, however, should be based on Ampere's circuital law, according to which the line integration of current-induced magnetic fields along a closed loop is equal to the current which intersects the closed loop. In order to perform this integration optically, we use Faraday effect, which states that the plane of linear polarization of a light beam propagating in an isotropic material subject to magnetization is rotated in proportion to the magnetization. So far many works on OCT have been made from engineering as well as scientific interests, and some OCTs developed have reached the stage of practical use and are commercially available. This chapter is devoted to the description of the present state-of-the-art of the OCT technology. The basic principles, types, performances and applications of OCTs are addressed.

2. THE BASIS

2.1. Optical Ampere's Circuital Law

When a steady electric current I flows in 3D space, according to the Ampere's circuital law,

$$I = \oint \vec{H} \cdot \vec{ds}, \tag{1}$$

where $\vec{H}$ is the current-induced magnetic field vector and $\vec{s}$ is the line vector along which the loop integration is carried out. In terms of Faraday effect, eq. (1) becomes

$$n\oint V\vec{H} \cdot \vec{ds} = nVI = \phi, \tag{2}$$

where V is Verdet constant of the loop material (assumed to be constant along the entire optical path), ϕ is Faraday rotation angle due to one round trip of light along the loop, and n is the number of turns of the optical loop. Thus, in order that the OCT based on eq. (2) may function properly, the optical loop has to be made of a homogeneous, isotropic and closed optical circuit (HICOC).

2.2. The HICOC Condition

Unless the HICOC condition is fulfilled, the OCT undergoes the measurement uncertainty such that the output signal of the OCT is affected by external currents of the measuring current, and also, by the relative position of the measuring current within the optical loop. In order to evaluate the measurement error due to imperfect fulfillment of the HICOC condition, we assume that the optical loop of an OCT involves some part which has a different Verdet constant V' from the other part, as shown in fig. 1(a). Faraday output signal then undergoes a measurement error due to the cross-talk with (or isolation from) a surrounding current as:

$$e = |(V'-V)/V|(\theta/2\pi), \tag{3}$$

where θ stands for the subtended angle from the current position to the path AB as shown in fig.1(a) [3,4]. Calculation of eq. (3) as a parameter of e is shown in fig. 1(b). We consider the case that the optical path is not closed but terminated at separated points A and B. This corresponds to $V' = 0$ on the AB line so that thereby-caused error due to cross-talk is [from eq. (3)] $e = \theta/2\pi$. In a particular case, in which the external current is located at a distance d from the AB line (AB $= L$), θ is then $\theta_1 = 2\tan^{-1}(L/2d)$. Therefore, referring to fig. 1(b), for $x = \theta_1/2\pi$ and $y = 1$, it follows that when d is, e.g. 10 mm, in order to achieve $e < 0.01$, L must be smaller than 0.63 mm. It is thus recognized that even the presence of such a slightly opened loop can cause a significant cross-talk error for close currents. On the other hand, when the traveling light has an elliptical polarization on the path AB, it corresponds to $V' = V \cos\Delta$ (Δ being the polarization retardation) so that $e < 0.01$ requires $\Delta \le 12°$.

Figure 2 shows the experimental demonstration of the cross-talk error due to an open loop [3,4]. In the sensor cell geometry shown in fig. 2(a), the optical path is definitely open without the (4 mm thick) compensating glass plate while the optical path is well-closed with it on the input/output port of the Faraday cell. For both cases, magnitudes of the cross-talk with an external current were measured as a function of the distance of the current from the center of the Faraday cell, and compared with each other in fig. 2(b), together with the theoretical values of eq. (3). Figure 2(b) clearly indicates that, in real OCTs, the cross-talk with close currents becomes a significant error if the optical path is not properly closed.

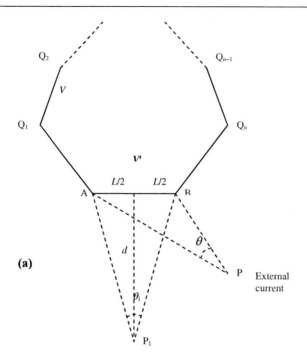

(a)

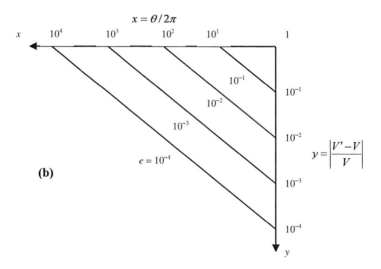

(b)

Figure 1. Theoretical value of cross-talk with surrounding current in OCT [3,4]; (a) geometry; (b) dependence of cross-talk e on subtended angle θ and Verdet constant inhomogeneity.

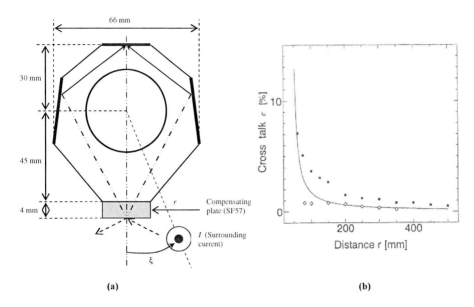

Figure 2. Demonstration of cross-talk with surrounding current in OCT [3,4]; (a) experimental arrangement, (b) dependence of cross-talk e on external current distance r at $\xi = 0°$ with (∘) and without (•) compensator. The solid line represents the calculation by eq. (3).

2.3. Signal Interrogating Schemes

Faraday signal can be detected by some different interrogating schemes. In general, a polarimetric interrogation, as follows, is employed. A linearly polarized light is incident on the Faraday material, the output light is received by two polarizers which are oriented at ±45° to the incident polarization direction, and the degree of polarization of the output light is detected as the OCT signal. As to the light source, a low coherent light is preferable to laser light to avoid its interference noises, and SLD (superluminescent laser diode) is especially used for the coupling to single mode fiber (SMF).

Because the Faraday rotation originates from the magnetically-induced circular birefringence of a substance, interferometric detection of the relevant optical phase shift yields the OCT signal. Though various types of interferometers (e.g., Fabry-Perot, Mach-Zender types) are, in principle, applicable to OCT, a ring interferometer (i.e., Sagnac interferometer) is the most suited one for the Faraday signal interrogation from a closed optical path in particular. The phase difference between the clockwise and the counter-clockwise traveling light beams in the interferometer yields twice the Faraday rotation [5], and then the undesirable reciprocal phase shifts due to temperature changes or vibration are cancelled. Similarly, a ring laser, or an active Sagnac interferometer, can also be used for the Faraday signal interrogation. Interference of the clockwise and the counter-clockwise traveling light beams going out of the ring laser generates an optical beat whose frequency is proportional to the Faraday rotation angle. The Faraday rotation angle is then interrogated in the frequency domain with a linear scale within the free spectral range of the laser [6,7].

In general, the Faraday signal detection is accompanied by optical noises. For dc current sensing, in particular, the drift of light power becomes a serious noise. To reduce it, the heterodyne detection using two frequencies of laser light [5,8] or a phase-modulated (or ac) Sagnac interferometer is very much suited.

3. ALL-FIBER TYPE OCT

An SMF made of silica is an ultra-low loss, thin, long, flexible, cheap and intrinsically isotropic optical waveguide, and therefore, its application to OCT started immediately after the invention of the fiber [9,10]. In reality, however, it took a long time to arrive at practical OCTs mainly because their development was much hindered by the involvement of linear birefringence (LB) in real fibers.

3.1. Magneto-Optical Properties of SMFs

The general theory for Faraday effect in isotropic SMFs is presented in ref. [11]. In the simultaneous presence of LB and Faraday effect in a medium, when a linear polarization is input into to medium along the principal axes of LB, the amplitude of the electric field of the output light is given by [12]

$$F = E_0 HVL \mathrm{sinC}\{[(\delta/2)^2 + VH]^{1/2}L\}, \qquad (4)$$

where δ is the linear retardation per meter, L is the optical path length, H is the applied magnetic field, E_0 is the amplitude of the electric field of the input light, and $\mathrm{sinC}x = \sin x/x$. Obviously, for $\delta = 0$, eq. (4) becomes $F = E_0\sin(VHL)$, which is the known relationship for Faraday effect in isotropic mediums. In the case that $|\delta| >> |VH|$, eq. (4) becomes $F = E_0VHL\mathrm{sinC}(\delta L/2)$, so that $|F| \leq 2E_0|VH/\delta|$. For silica fibers subjected to a bending of e.g. 50 mm radius of curvature, $|\delta|$

amounts to the order of 1 radian/m while $|VH|$ is typically 10^{-5} I rad.$A^{-1}m^{-1}$; it then follows that $|HV/\delta| \le 10^{-5}$ I A^{-1}, and hence, the field conversion efficiency $|F|/E_0$ is smaller than the order of 10^{-5} I A^{-1}, which is usually very small except for especially heavy currents.

Therefore, some approaches to eliminate LB in silica fibers were conducted. One of them is to induce circular birefringence (CB) by providing a post-drawing fiber with twist [13,14] or an in-drawing fiber with spin (i.e., spun fiber) [15]. For such a fiber, eq. (4) becomes

$$F = E_0 H(VH + \delta_T / 2)L \text{sinC}\{[(\delta/2)^2 + (VH + \delta_T / 2)^2]^{1/2} L\}, \quad (5)$$

where δ_T is the provided CB/m. In the case when CB is much larger than LB (i.e., $|\delta_T| >> |\delta|$), eq. (5) becomes

$$F = E_0 \sin[(VH + \delta_T / 2)L], \quad (6)$$

which indicates that the effect of LB is neglected against CB, and F becomes the same as that for the fiber without LB case except involving a constant bias polarization rotation. When a twist rate of 1 turn/m is applied to a post-drawing silica SMF, $|\delta_T|$ amounts to 1 rad/m.

The study of the thermal relaxation of fiber stress has been attempted, and it has been shown that the bend-induced LB in silica fiber could be reduced by annealing [16]. The annealed fiber, however, is still sensitive to newly applied stress so that, in its practical use, a careful remedy is needed to avoid the application of new stress.

Naturally, it is most desirable to invent an SMF which has inherently no photo-elastic effect. Such fiber was developed by using flint glass as the fiber material [17]. Compared with silica, flint glass has about 100-times smaller photo-elastic constant (Brewster coefficient) and, of additional advantage, its Verdet constant is 6-times larger ($V = 1.5 \times 10^{-5}$ rad/A at a wavelength of $\lambda = 850$ nm) and little temperature dependent. The transmission loss is, however, pretty high. The flint glass SMF is now commercially available (HOYA Co., Tokyo, Japan).

One more point to be noted in the application of SMF for OCT sensors is the geometrical rotation of the polarization plane of the transmitted light from the fiber. As a general feature of light transmission in optical fibers, if the fiber is not in-plane, the light is then subjected to a purely geometrical rotation of the polarization plane, depending on the geometry of the fiber path [18]. In all-fiber

OCTs, this geometrical polarization rotation can cause polarization instability. Fortunately, this geometrical polarization rotation is a reciprocal effect so that it can be cancelled by making the light beam to go forth and back in the same fiber (i.e., by the use of the reflection scheme) [19]. In the transmission type OCTs, however, this geometrical rotation can introduce uncertainty in measurement due to polarization instability. As such, the sensing fiber coil needs to be fixed on a rigid plane in a stable manner.

3.2. Practical All-Fiber OCTs

Various types of all-fiber OCTs have been developed so far and several of them have been put into either test or practice for the electric power systems including power generators in power plants, railway systems, nuclear devices etc.

Regarding flint-glass fiber OCTs, both the transmission and reflection types with polarimetric interrogation have been developed [20]. The transmission-type, in which a flint glass fiber coil is fixed on a durable metallic film, could measure currents up to 300 kA (which corresponds to a Faraday rotation of 45°) within a measurement error of 2 kA to meet the Japanese industrial standard [20]. Figure 3(a) shows the optical diagram of the reflection-type OCT. A low coherent light from SLD (λ = 1550 nm) is remotely sent (via a polarization maintaining fiber) to the sensing part, and coupled into the flint glass fiber after being passed through an optical unit that serves as both a polarizer and a polarization bias element. The reflected light from the sensing fiber is again passed through the same optical unit and ±45° polarization components of the emerging light are received (via an SMF) by two photo detectors 1 and 2. Figure 3(b) shows the application of the reflection-type OCT for fault current detection. The sensing fiber could, on account of the reflection-type, freely encircle the three phases of electric cables, and accurate phase current detection (summation of the three phase currents) is carried out [21].

The annealed-silica-fiber OCTs with polarmetric interrogation have been put into for testing. In order to reduce the effects of thermal expansion and vibration, the annealed fiber coil is installed in the glass epoxy compound package [22]. A Sagnac interferometer OCT using annealed silica SMF has also been tested for applications, and such OCTs are found to have a measurement accuracy of better than 0.5% for the environmental temperature range from −35 °C to 85 °C [23]. A Sagnac interferometer OCT using spun silica SMF has also been developed for the application in D.C. railway transportation system [24].

All of the all-fiber OCTs reported so far, unfortunately, did not describe about the closing feature of the sensing fiber loop so that there are ambiguities about the possible cross-talk errors with very close surrounding currents.

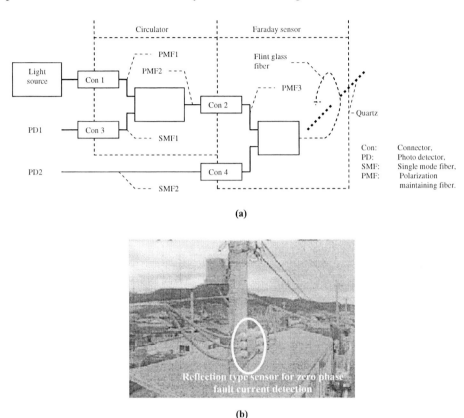

(a)

(b)

Figure 3. Practical flint glass OCT of the reflection type [21] ; (a) optical diagram, (b) fault section locating system for 22 kV power cable lines.

4. BULK-OPTIC OCT

In order to get rid of the problem of fiber LB or to make it possible to use a variety of Faraday materials, it is important to study the use of non-fiber bulk-optic materials as the Faraday element. In this type of OCT (or the bulk-optic type of OCT), the light beam travels in a zigzag path within a Faraday cell. The key requirement for the bulk-optic OCT is to achieve the polarization-maintaining

100% reflectance of the traveling light at each reflecting surface of the Faraday cell. To this end, the oldest scheme applied the crossed right-angle total-reflection glass prisms for the corner reflection [25,26] but it, in principle, could not well fulfill the HICOC condition. In order to fulfill the HICOC condition precisely, new schemes have been developed, which applied the thin-film coating technology. Two different approaches have been conducted. The first approach extended the well-known quarter-wavelength multilayer coating technology to perform the polarization-maintaining 100% reflectance. The theory required the specific conditions of "above Brewster's angle and below critical-angle" of light on every boundary of the multilayer thin films. The theoretical requirement was experimentally demonstrated by the use of conventional coating materials. The HICOC condition was thus for the first time very exactly fulfilled [4]. The resulting form of the Faraday cell was a polygon of specific form.

The second approach used the coating of the bi-layer dielectric thin films on the total reflection surface to achieve the polarization-maintaining 100% reflectance for light beam at oblique angle of incidence [27]. The optical design was conducted by the trial-and-error numerical calculation. The sensor system developed is shown in fig. 4(a) [27]. Each of the three reflecting surfaces of a square block Faraday cell made of flint glass was coated by the suitably thick SiO_2 and Ta_2O_5 films to produce no polarization-retardation and 100% reflectance on the internal reflection for 45° oblique incidence of light. The input/output port of the Faraday cell was provided by attaching a low-Verdet-constant (SF7) tiny prism on one uncoated surface of the cell. This Faraday cell had good tolerances with respect to light beam angle, wavelength and film thickness. The current sensing characteristics of this OCT are shown in fig. 4(b). The polarization extinction ratio was better than 2×10^{-4} and high isolation from the surrounding current was demonstrated [fig. 4(c)]. Figure 4(d) shows the ac Faraday signal and the dc light power as a function of applied weight to the cell; it indicates that the OCT is very stable against the mechanical disturbances applied to the Faraday cell.

5. HYBRID TYPE OCT

The third type of OCT is fabricated by a simple modification of the traditional electric CT. This type of OCT, or the hybrid OCT, is composed of a high-permeability (metallic) magnetic core with a small gap, and a Faraday element

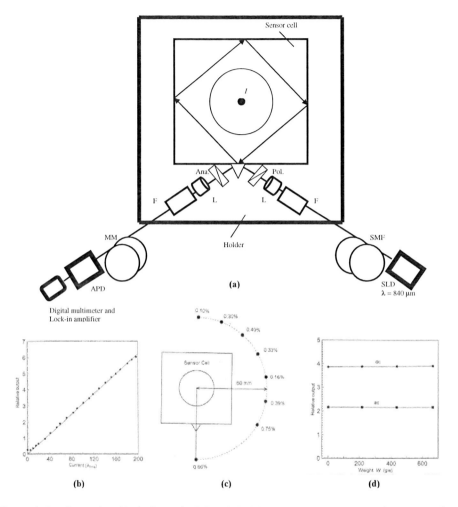

Figure 4. Precise and stable bulk-optic OCT [27]; (a) current sensor system, (b) measured current characteristics, (c) measured cross-talk at different positions of the external current, (d) measured weight characteristics of dc and ac Faraday signals, indicating high mechanical stability of the OCT.

inserted in the core gap. In principle, the hybrid type OCT cannot rigorously fulfill Ampere's circuital law, and therefore, cannot be a much exact OCT. Moreover, the use of metallic magnetic cores as the field concentrator reduces the insulating feature of OCT, and saturates the Faraday signal against large currents. Despite these drawbacks, the hybrid type OCT is an attractive one because,

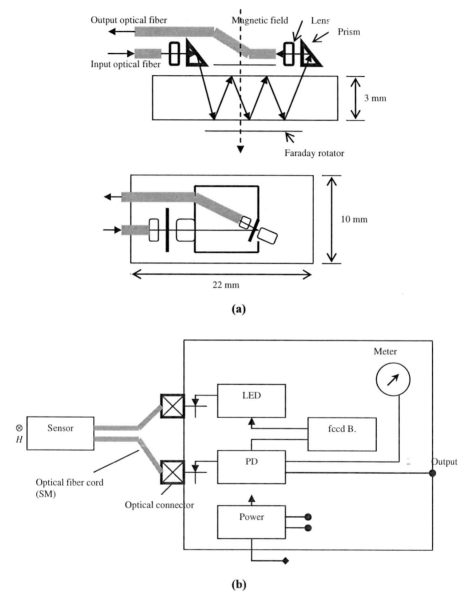

Figure 5. Optical magnetic-field sensor of the multiple-reflection type [30]; (a) sensor head, (b) instrument system.

compared with the bulk-optic type OCT, the hybrid one allows the use of various Faraday materials, which include crystals and thin films, and does not need specific internal refection schemes of the Faraday cell (unlike the bulk-optic

OCT). Also, the hybrid type OCT can be used as a clamp-type OCT, like the electric CT. In order to achieve high isolation from surrounding currents and to improve the accuracy of OCT, it is required that the gap length of the core is possibly small. This is the key requirement for developing high performance hybrid type OCT. Earlier studies developed the multiple-reflection Faraday cell made of a thin FR5 glass plate, as shown in fig. 5, which is suitable for the insertion into a small gap of a ring core [28–30].

An especially attractive application of the hybrid type OCT is the small current measurement, which is very difficult with the all-fiber or bulk-optic type OCT. The achievable sensitivity is governed by the performances of Faraday materials and magnetic core systems.

As for high Verdet constant materials, transparent magnetic materials such as rare earth iron garnet (RIG), e.g., yttrium iron garnet (YIG), are attractive. Improvements on their magneto-optical properties such as temperature characteristics [31], linearity [32] and response speed [33] were conducted. Especially, as for the temperature characteristics, Bi-doped YIG with high temperature stability has been developed, and is now commercially available.

A magnetic material (MM) has the following inherent features as the Faraday material:

(i)   Verdet constant is typically 100 times higher than non-magnetic materials,

(ii) when MM is applied in the hybrid OCT, because of its high permeability, the total magnetic resistance of the core system becomes small, thereby enhancing the current-sensitivity,

(iii) it is possible to magnetize MM in the other directions than that of the applied magnetic field by virtue of demagnetization and/or magnetic-anisotropy (such as uniaxial anisotropy) effects peculiar to MM,

(iv) the magnetic response of MM is subjected to nonlinearity, hysteresis and saturation against the applied magnetic field, unlike non-magnetic materials,

(v)   the produced magnetization is not uniform in MM because of the domain structure, which scatters or diffracts the light beam, and hence, causes the measurement uncertainty. This domain effect, however, is averaged and can be reduced by the illumination of a broad-width light beam as the probe light.

It results from the above features of MM that a possibly smaller core gap and a broad-beam illumination are desirable for the hybrid OCT using MM. Usually, in the core gap, the probe light is passed through the Faraday element in the direction parallel to the core gap (longitudinal configuration). Then, because the gap length is very limited, special optical designs such as a core system with holes [34] or a micro-optic system [35] have been introduced.

It has also been demonstrated that the hybrid OCT using MM functions well even in the transverse configuration where the probe light is passed through the Faraday element in the direction perpendicular to the core gap [36,37]. The operating principle of this unusual hybrid OCTs is particularly based on the feature no. (v) of MM as mentioned above.

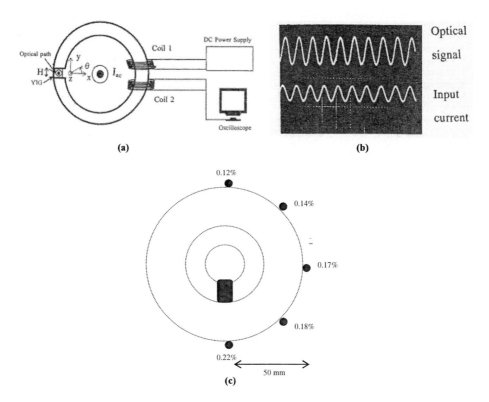

Figure 6. Fiber-linked YIG/ring-core Faraday effect optical current sensor in the transverse configuration [36]; (a) entire system, (b) oscilloscope traces of optical signal for 50Hz-100mA$_{rms}$ current, (c) measured cross-talk from the surrounding current at various angular positions.

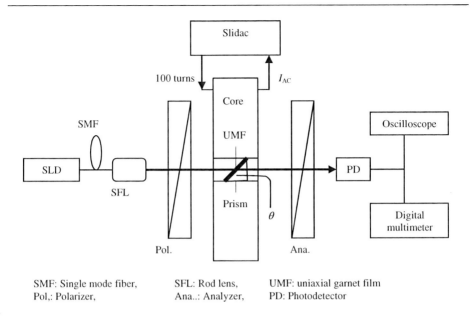

SMF: Single mode fiber,   SFL: Rod lens,   UMF: uniaxial garnet film
Pol,: Polarizer,   Ana..: Analyzer,   PD: Photodetector

Figure 7. Schematic diagram for magnetic garnet film/ring core Faraday effect optical current sensor in the transverse configuration [37].

Figure 6(a) shows the entire system of the hybrid OCT using a YIG bulk crystal Faraday cell in the transverse configuration [36].

The YIG crystal ($3\times10\times10$ mm^3) was tightly inserted in a small gap (of 3 mm) of a permalloy core (of the size 40 mm outer diameter $\times$ 10 mm thickness) to overcome the problem of air gap. Under the illumination of a light beam (from 1.3 μm SLD) on the crystal in the transverse direction to the core, a very high current sensitivity of a modulation depth of 20% per Ampere was obtained, which has been the highest sensitivity of OCT ever reported and about 10^4 times higher than that of the flint-glass fiber OCT. Figure 6(b) shows the typical optical response signal. Good linearity for currents from 0 to 1 Ampere was obtained and the isolation from surrounding currents was very high, as shown in fig. 6(c).

Figure 7 shows the entire system of the hybrid type OCT using a uniaxial Bi-substituted garnet thin film (650 μm thick) in the transverse configuration [37]. The garnet film was obliquely (typically at 45° inclination angle) inserted into the 3.5 mm air gap of a permalloy ring core, and was illuminated by a collimated optical beam (1.3 μm SLD) from the transverse side of the core. The high sensitivity of about 0.08% per Ampere modulation-depth, good linear response from 0 to 1 Ampere and good reproducibility were demonstrated.

6. OTHER OCTs AND REMARKS

The OCT technology is closely related to the optical magnetic field sensing technology. Our descriptions are limited to the OCT which basically relies on the Ampere's circuital law for the sensing principle. However, if there are no other currents than a measuring current, and moreover, the magnetic field distribution due to the measuring current is a priori known, then other techniques than the ones based on the optical Ampere's law can also be applied for the current sensing. One of the useful techniques for this purpose is the introduction of a solenoid coil to transform a current into a uniform magnetic field within the solenoid coil, as long as the resistance and reactance of the solenoid do not matter. Various types of solenoid-combined OCTs were studied in connection with the optical sensing of electric power, and their actual performances are described in refs. [38–41].

CONCLUSION

The OCT based on Faraday effect has the invaluable features of non-contact, highly insulating and high-speed measurements. Studies on OCT have been made to fulfill the HICOC condition satisfactorily. The currently-known most effective HICOC technologies use ultra-low photo-elastic coefficient fibers, dielectric-thin film coatings and magnetic Faraday materials in the transverse configuration, for the all-fiber, bulk-optic and hybrid type of OCTs, respectively. Such OCTs can provide very simple and flexible sensor head, accomplish the fulfillment of the HICOC condition very precisely and achieve very high current sensitivity. OCTs are now mostly applied for the electric power industries, but can find other wide applications too in both engineering and science. It is very much expected that the OCT technology will improve more with the progress of optoelectronic materials and devices.

REFERENCES

[1]   Yoshino, T. *Proc. SPIE* 1987, 798, 258–266.
[2]   Day, G.W.; Rose, A.H. *Proc. SPIE* 1988, 985, 138–150.
[3]   Yoshino, T.; Takahashi, Y.; Gojyuki, M. *Opt. Rev.* 1997, 4, 108–110.
[4]   Yoshino, T.; Gojyuki, M.; Takahashi, Y.; Shimoyama, T. *Appl. Opt.* 1997, 36, 5566–5573.

[5]   Arditry, H.J.; Bourbin, Y.; Mapuchon, M.; Puech, C. *Tech. Dig. of the 3ʳᵈ Int. Conf. on Integrated Optics and Optical Fiber Communication (IOOC'3, San Francisco)* 1981, 128–129.

[6]   Kueng, A.; Nicati, P.A.; Robert, P.A. *Opt. Rev.* 1997, 4, 56–57.

[7]   Takahashi, Y.; Yoshino, T. *J. Light. Tech.* 1999, LT-17, 591–597.

[8]   Yoshino, T.; Hashimoto, T.; Nara, M.; Kurosawa, K. *J. Light. Tech.* 1992, LT-10, 503–513.

[9]   Smith, A.M. *Appl. Opt.* 1978, 17, 52–56.

[10] Rapp, A.; Harms, H. *Appl. Opt.*1980, 19, 3729–3747.

[11] Yoshino, T. *J. Opt. Soc. Am. B* 2005, 22, 1856–1860.

[12] Yoshino, T. (ed.) Optical fiber sensor technologies (2ⁿᵈ ed.), Dai-ichi International Co.: Tokyo, 1986, 15–160.

[13] Ulrich, R.; Simon, A. *Appl. Opt.* 1979, 18, 2241–2251.

[14] Rashleigh, S.C.; Ulrich, R. *Appl. Phys. Lett.* 1979, 34, 769–770.

[15] Laming, R.I.; Payne, D.N.; *J. Light. Tech.* 1989, 7, 2084–2094.

[16] Tang, D.; Rose, A.H.; Day, G.W., Etzel, S.M. *J. Light. Tech.* 1991, 9, 1031–1037.

[17] Kurosawa, K.; Yoshida, S.; Sakamoto, K. *J. Light. Tech.* 1995, 13, 1378–1384.

[18] Ross, J.N. *Opt. Qunatum Electron.* 1984, 16, 455–461.

[19] Yoshino, T.; Iwama, M. *Opt. Lett.* 1999, 24, 1626–628.

[20] Kurosawa, K. *Opt. Rev.* 1997, 4, 38–44.

[21] Kurosawa, K.; Shirakawa, K.; Saito, H.; Itakura, E.; Sowa, T.; Hiroki, Y. Kojima, T. *Proc. of the 16ᵗʰ Int. Conf. on Optical Fiber Sensors* (OFS'16, Nara) 2003, 316–319.

[22] Willsch, M.; Bosselmann, T. *Proc. of the 15ᵗʰ Int. Conf. on Optical Fiber Sensors* (OFS'15, Boulder) 2002, 407–410.

[23] Bohnert, K., Gabus, P.; Brandle, H., *Proc. of the 16ᵗʰ Int. Conf. on Optical Fiber Sensors* (OFS'16, Nara) 2003, 752–755.

[24] Hayashiya, H.; Kumagai, T.; Hino, M.; Endo, T.; Ando, M.; Negishi, H. *Proc. of the 32ⁿᵈ Meeting on Lightwave Sensing Technology (Jpn. Soc. Appl. Phys., LST'32, Tokyo)* 2003, 141–146.

[25] Saito, S.; Fujii,Y.; Yokoyama, K.; Hamasaki, J.; Ohno, Y. *IEEE J. Quantum Electron.* 1968, QE-2, 255–259.

[26] Kanai, N.; Takahashi, G.; Sato, T.; Higashi, M.; Okamura, K. *IEEE Trans. Power Del.,* 1986, PWRD-1, 91–97.

[27] Yoshino, T.; Yokota, M.; Aoki, K.; Yamamoto, K.; Itoi, S.; Ohtaka, M. *Appl. Opt.* 2002, 41, 5963–5968.
[28] Yoshino, T. *Jpn. J. Appl. Phys.* 1978, 19, 745–749.
[29] Yoshino, T.; Ohno, Y. *Fiber and Integrated Opt.* 1981, 3, 391–399.
[30] Yoshino, T.; Ohno, Y.; Kurosawa, K. *Tech. Dig. of the 2nd Int. Conf. on Optical Fiber Sensors (OFS'2, Stuttgart)* 1984, 55–58.
[31] Inoue, N.; Yamasawa, K. T. *IEE Japan* 1995, 115-A, 1114–1120.
[32] Numata, T.; Tanakaike, H.; Inokuchi, S.; Sakurai, S. *IEEE J. Mag.* 1990, 26, 1358–1360.
[33] Rochford, K.B.; Rose, A.H.; Deeter, M.N.; Day, G.W. *Opt. Lett.* 1994, 19, 1903–1905.
[34] Yoshino, T.; Hara, H.; Sakamoto, N. *Extended Abstracts of the Spring Meeting of Jpn. Soc. Appl. Phys.*, 1988, 860.
[35] Itoh, N.; Minemoto, H.; Ishiko, D.; Ishizuka, S. *Tech. Dig. of the 11th Int. Conf. on Optical Fiber Sensors (OFS'11, Sapporo)* 1996, 638–641.
[36] Yoshino, T.; Minegishi, K.; Nitta, M. *Meas. Sci. Tech.* 2001, 12, 850–853.
[37] Yoshino, T.; Torihata, S.; Yokota, M.; Tsukada, N. *Appl. Opt.* 2003, 42, 1769–1772.
[38] Li, Y.; Li, C.; Yoshino, T. *Appl. Opt.* 2001, 40, 5738–5741.
[39] Li, C.; Cui, X.; Yoshino, T. *IEEE Tans. Instrum. Meas.* 2001, 50, 1375–1380.
[40] Li, C.; Yoshino, T. *Appl. Opt.* 2002, 41, 5391–5397.
[41] Li, C.; Cui, X.; Yoshino, T. *J. Light. Tech.* 2003, 20, 843–849.

In: New Developments in Optics Research
Editor: Matthew P. Germanno

ISBN: 978-1-60324-505-7
© 2012 Nova Science Publishers, Inc

Chapter 2

EVANESCENT FIELD TAPERED FIBER OPTIC BIOSENSORS (TFOBS): FABRICATION, ANTIBODY IMMOBILIZATION AND DETECTION

Angela Leung[1], P. Mohana Shankar[2] and Raj Mutharasan[1]*

[1]Department of Chemical and Biological Engineering
[2]Department of Electrical and Computer Engineering, Drexel University, Philadelphia, PA, U. S.

ABSTRACT

Tapered Fiber Optic Biosensors (TFOBS) are sensors that operate based on fluctuations in the evanescent field in the tapered region. In our laboratory, TFOBS are made by heat pulling commercially-available single mode optical fibers. They have been investigated for various applications, including measurement of physical characteristics (refractive index, temperature, pressure, etc.), chemical concentrations, and biomolecule detection. In this chapter, an up-to-date review of TFOBS research is provided, with emphasis on applications in biosensing such as pathogen, proteins, and DNA detection. The physics of sensing and optical behavior based on taper geometry is discussed. Methods of fabrication, antibody immobilization, sample preparation, and detection from our laboratory are

---

[*] E-mail address: mutharr@drexel.edu. Tel.: (215) 895-2236. Fax: (215) 895-5837. (Corresponding author)

described. We present results on the non-specific response, simulation, and detection of *E.coli O157:H7* and BSA. This chapter will conclude with an analysis of the future direction of the Tapered Fiber Optic Biosensors.

1.0. INTRODUCTION

Tapered fiber optic biosensors (TFOBS) are made from optical fibers, and, are capable of detecting specific analytes using optical responses. They have been used for the measurement of physical and chemical properties [4-8], [9-14] of biological molecules [2, 15-19] and have several applications including environmental monitoring, drug screening, clinical diagnostics, and defense. TFOBS offer many advantages including flexibility, ease of use, affordability, and ability to perform sensing using a small amount of sample. These sensors are based on the evanescent field associated with fiber, and, often are also referred to as Evanescent Field Tapered Fiber Sensors. In this chapter, the basics of TFOBS are discussed, along with an up-to-date literature review of TFOBS. Experimental methods and recent results from our laboratory are also presented.

2.0. PHYSICS OF EVANESCENT FIELD SENSING IN TAPERED FIBERS

Optical fibers are cylindrical waveguides, and, are made of a silica core surrounded by a silica cladding. The core refractive index is higher than the cladding refractive index (RI) because it is doped with Ge. Light propagates through the core by total internal reflection (TIR). Besides the light propagating in the core, there is a small component of light, known as the evanescent field, which decays into the cladding.

Evanescent light penetration is described by its penetration depth (d_p), which is the position away from the core/cladding interface at which the light decays to $1/e$ of its value at the core-cladding interface, and is given by:

$$d_p = \frac{\lambda}{2\pi\sqrt{n_{co}^2 \sin^2\theta - n_{cl}^2}} \tag{1}$$

In eqn. (1), λ is the operating wavelength of light, n_{co} the index of the core and n_{cl} the index of the cladding. The angle of incidence at the core cladding interface is θ. The evanescent field in a uniform diameter fiber does not interact with the outside environment because it decays to a negligible value as it reaches beyond the cladding. This is due to the fact that in typical fibers the cladding thickness is several times that of the core. However, if the cladding is removed or the fiber is tapered down to a diameter less than the original core diameter, evanescent field can interact with the external medium affecting the transmission through the fiber.

The penetration depth in a tapered fiber depends on the local diameter of the tapered fiber, the RI of the core, and RI of the external medium. Since there is a continuous change in the diameter along the fiber in the tapered region (except in the waist), coupling of light among the modes can occur [20]. Coupling in the tapered region causes the transmission properties of the fiber to change. Presence of analytes in the tapered region can lead to RI changes in the taper. This results in changes in the coupling characteristics and causes changes in the optical throughput.

Physical characteristics of the fiber such as RI of the core and cladding, core diameter, and operating wavelength determine the number and type of modes that propagate through the fiber. The lowest order mode has the tightest confinement of the field, and hence the weakest evanescent field. As the mode order goes up, the associated evanescent field also increases. In a tapered or de-cladded fiber, the optical characteristics of the surrounding medium such as its index, absorption, etc. can affect the optical throughput.

2.1. Wave Propagation in Absorption Sensors

The shape of the optical field in a fiber is determined by the number of modes present. Figure 1 shows a tapered fiber with a short region of constant thickness (waist) and contracting and expanding regions. The number of modes that can be supported in a fiber is determined by the V-number,

$$V = \frac{2\pi a_0}{\lambda}\sqrt{n_{co}^2 - n_{cl}^2} \qquad (2)$$

where a_0 is the radius of the core. When $V \leq 2.405$, only the lowest order mode is supported, and as V increases, number of modes increase. Although in a single-mode (SM) fiber only the lowest order is supported, in the tapered region higher order modes can potentially be supported because of the larger difference in refractive index between the core and the sample (~0.12) compared to a regular fiber (~0.01). Reduction of the fiber radius increases the evanescent field strength, and enhances the interaction of the evanescent field with the analyte leading to variations in optical throughput (transmitted light).

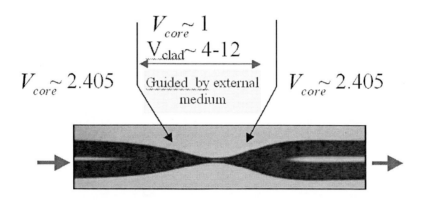

Figure 1. Photograph of a TFOBS. The region of interest in a tapered fiber is identified by the region where V_{core}<1.

2.2. Wave Propagation in Continuous Bi-conical Tapered Fibers

In our laboratory, tapered fibers were made by heat-pulling an optical fiber without removing the cladding. Unlike uniform fibers, the V-number changes along the length of a tapered fiber. When V-number becomes less than unity, the core is too small to contain the light and light guidance is determined by the original cladding which acts as the core and the external medium of RI n_{ext} which serves as the cladding. The new V-number is called V_{clad} where the parameter, a_0 in Eq. (2) is replaced by the radius of the overall fiber, b(z), and is given by

$$V_{clad}(z) = \frac{2\pi b(z)}{\lambda}\sqrt{n_{cl}^{2} - n_{ext}^{2}} \tag{3}$$

In tapered fibers, the V-value is generally referred to as V_{clad} to distinguish it from V_{core}, given in Eq. (2). Note that the diameter b in eqn. (3) is a function of the location (z), indicating the existence of tapering.

2.3. Numerical Simulation of Light Transmission in a Tapered Fiber

Numerical simulation of light transmission through a tapered fiber can provide useful insight into its properties. To simplify the analyses, the simplified mode theory based on linearly polarized modes (LP) can be used to determine the transmission behavior [21]. Assuming that light enters into the fiber parallel to the axis, the only modes that are excited are the LP_{0m} modes. The transverse components of the electrical field inside the fiber are:

$$E_X(r) = A \begin{cases} J_0(U_{0m}R), & R \leq 1 \\ \dfrac{J_0(U_0)}{K_0(W_0)} K_0(W_{0m}R), & R > 1 \end{cases}$$ (3)

where R is the normalized radial coordinate, r/a_0, whereas U and W are constants They depend on the wavelength and RI of the core and cladding,

$$U_{0m}^2 = a_0^2 \left[\left(\frac{2\pi}{\lambda} \right)^2 n_{co}^2 - \beta_{0m}^2 \right]$$ (4a)

$$W_{0m}^2 = a_0^2 \left[\beta_{0m}^2 - \left(\frac{2\pi}{\lambda} \right)^2 n_{cl}^2 \right]$$ (4b)

where c is the speed of light and β is the propagation constant. When m=1, only the fundamental mode exists. In eqn. (3) $J_0(.)$ and $K_0(.)$ are the Bessel and modified Bessel functions of zero[th] order, respectively. A is a constant determined from orthogonality principle [21]. The subscript m represents the various circularly symmetric LP_{0m} modes that may be present in the fiber.

When $V_{core}<1$ and $V_{clad}>2.405$, many modes are supported since the index difference between the cladding and the external medium (n_{ext}) is large. As mentioned previously, tapering leads to coupling among LP_{0m} modes [20, 22] . A simple means to visualize the taper is to model taper geometry by approximating the slopes of the taper by stepwise linear approximation. At each step 'i', the parameters U_{0m} and W_{0m} are analogous to the constants U and W in a uniform diameter fiber. They can be expressed using the local radius ρ^i as

$$U_{0m}^{i\,2} = \rho^{i\,2}\left[\left(\frac{2\pi}{\lambda}\right)^2 n_{cl}^{\,2} - \beta_{0m}^{i\,2}\right] \tag{5a}$$

$$W_{0m}^{i\,2} = \rho^{i\,2}\left[\beta_{0m}^{i\,2} - \left(\frac{2\pi}{\lambda}\right)^2 n_{ext}^{\,2}\right] \tag{5b}$$

The V-number for each step is given by:

$$V^i = \frac{2\pi}{\lambda}\rho^i\left[n_{cl}^{\,2} - n_{ext}^{\,2}\right]^{1/2} \tag{6}$$

The values of U, W and β are calculated using the LP mode approximation [21]. The relationship between the modal amplitudes of the LP_{0m} modes of the ith and $(i+1)^{th}$ step is:

$$\sum_{m=1} A_m^i E_m^i(r)e^{-j\beta_m^i z^i} = \sum_{q=1} B_q^{i+1} E_q^{i+1}(r)e^{-j\alpha_q^{i+1} z^{i+1}} \tag{7}$$

where $E(r)$ is the electric field, β_m is the propagation constant on the left and α_q is the propagation constant on the right. It has been assumed that $E(r)$ are orthonormal [21]. A_m is the amplitude of the modes on the left and B_q is the amplitude of the modes on the right. That is,

$$\int_0^\infty \int_0^{2\pi} |E(r)|^2 r\,dr\,d\phi = 1 \tag{8}$$

The amplitude on the right is obtained by applying the orthogonality principle:

$$B_q^{i+1} e^{-j\alpha_q^{i+1} z^{i+1}} = \sum_n \sum_m A_m^i e^{-j\beta_m^i z^i} C_{nm;pq} \tag{9}$$

$$C_{nm;pq} = 2\pi \int_0^\infty E_q^{i+1}(r) E_m^i(r) r\,dr \tag{10}$$

In the tapered region, light is coupled among the various LP_{0m} modes. When $V_{core}=1$, power in the LP_{01} cladding mode is transferred to LP_{01} core mode and appears at the output end of the fiber. The light remaining in other modes stays in the cladding and is lost. A MATLAB® program was used to estimate the amplitude and output power. The taper geometry, wavelength and number of steps were varied to determine the resulting changes in power.

Sample simulation results, illustrated in Figure 2, show changes in transmission vs. waist diameters for two taper geometries. In Panel A, the taper geometry resembles a symmetric taper made by the fusion splicer, while in Panel B, the simulation is for a long taper similar to a heat-drawn taper. The transmission is normalized with respect to air, so that a value of 1.2 indicates a transmission increase of 20% in water compared to air. Figure 2 show that as the waist radius increases, the difference in transmission between water and air decreases. However, at intermediate values the ratio may be higher or lower than unity, particularly at smaller waist diameters. For certain values of the radius, the transmission through water is higher than in air. At a longer wavelength (550 nm, for example) and for the same waist diameter values, the difference in transmission between air and water differ by less than 10%. To explore further, simulation was undertaken by varying the diameter of the waist in much smaller steps of 0.001 μm. These results are shown in Figure 3 for two starting diameters, 5 μm and 6.25 μm. The transmission characteristics change significantly for small changes in diameter. For example, a 5.54 μm diameter taper exhibits 30% higher transmission in water while a 5.58 μm waist diameter taper transmits 20% less transmission. The differences, however, become smaller for larger diameters. The example of 6.25 μm in Figure 3 shows that the changes in transmission were less than 10 %.

These simulation results can serve as a guide to the analysis and interpretation of the experimental data on the tapered fibers. It is important to recognize that in the simulation we considered only the effect of refractive index in the waist region. In actual sensing experiments, the cells absorb at the operating wavelength, and the resulting sensor response is a complex interplay of these two phenomena. Furthermore, cells do not have homogeneous RI because the cells constitute particulate matter. Finally, the cell attachment onto the taper surface is often not uniform as we showed in our earlier report [2].

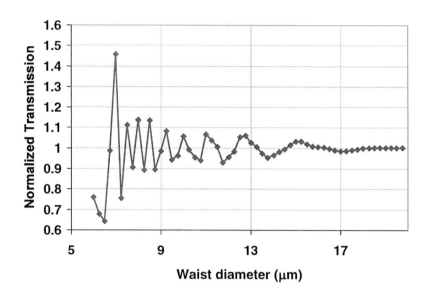

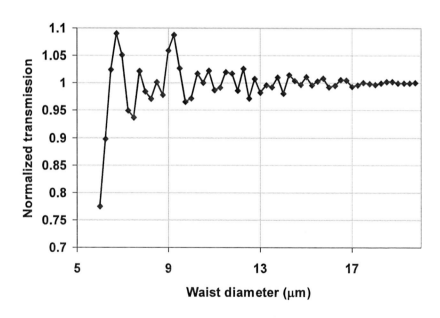

Figure 2. Transmission in water normalized with respect to air as waist diameter is altered. Top: A short symmetric taper: a= 0.425 mm, b=0.325 mm, c=0.500 mm. Operating wavelength = 470 nm. Bottom: A longer asymmetric taper: a=2.25 mm, b=0.245 mm, c=4.5 mm. Operating wavelength = 550 nm. Adapted from [3].

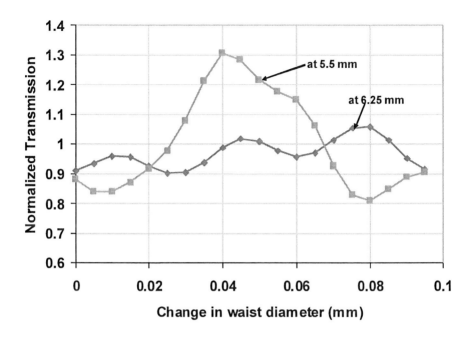

Figure 3. Transmission characteristics of a taper (a= 2.25 mm, b=0.25 mm, c=4.5 mm) in water at 470 nm as a function of change in waist diameter. Transmission is normalized with respect to transmission in air at the corresponding geometric values. Smaller starting diameter tapers show large changes in transmission for small (0.01 μm) changes is waist diameter. Adapted from [3].

3.0. LITERATURE REVIEW

In this section, the applications of TFOBS for pathogen detection, toxins measurements, clinical measurements, and DNA detection are presented. In tables 1 to 3, we summarize the analytes detected, matrices in which they were detected, detection principle, basis of sensors, and detection limits.

Table 1. TFOBS for Pathogen and Toxin Measurement

Target Analyte	LOD	Matrix	Taper Geometry	Fiber Type	Detection Principle	References
Bacillus anthracis	3.2E5 spores/mL	buffer	BT	Polystyrene MM	Fluorescent sandwich assay	[32]
Bacillus subtilis var. niger	8 x 10(4) spores/mL	buffer	NA (chip)	NA (chip)	Leaky wave (SPR)	[59]
LacZ DNA in Escherichia coli	25 pM	buffer	Uniform	MM	Fluorescent intercalating agents	[55]
Staphylococcus aureus Protein A	1 ng/mL	ND	ND	MM plastic	Fluorescent sandwich assay	[35]
Escherichia coli O157:H7	0.016 dB/h/N_0, Initial number (N_0): 10-800 *	buffer	BT	MM	Absorption	[23]
Escherichia coli O157:H7	70 cells/mL	Buffer	BT	SM	Intensity	[2]
Escherichia coli O157:H7	1 CFU/ml	ground beef samples	Uniform	MM polystyrene	Fluorescent sandwich assay	[25, 26]
Salmonella	50 CFU/g	irrigation water used in the sprouting of seeds	RAPTOR – uniform	Waveguide	Fluorescent sandwich assay	[27]
Salmonella	10(4) CFU/ml	Hotdog samples	RAPTOR – uniform	Waveguide	Fluorescent sandwich assay	[60]
Salmonella	10(4) CFU/mL	Nutrient broth	TT	MM	Fluorescent sandwich assay	[28]

Target Analyte	LOD	Matrix	Taper Geometry	Fiber Type	Detection Principle	References
Staphylococcal Enterotoxin B	Min: 0.5 ng/ml (buffer)	Buffer, human serum, urine, and aqueous extract of ham	CTT	ND	Fluorescent sandwich assay	[33]
C. Botulinum toxin A, Pseudexin Toxin	Min: 30 pM (*C. Botulinum* toxin A), 60 pM (Pseudexin)	ND	CTT	MM	Fluorescent sandwich assay	[61]
Clostridium-Botulinum Toxin-A	5 ng/mL	buffer	TT	MM	Fluorescent sandwich assay	[36]
E. coli lipopolysaccharide endotoxin	Min: 10 ng/ml	Buffer and plasma	CTT	ND	Fluorescent sandwich assay	[38]
Ricin Concentration	Min: 100 pg/ml (buffer) Max: 1 ng/mL (river water)	Buffer, river water	CTT	MM (plastic clad silica)	Fluorescent sandwich assay	[37]
Listeria monocytogenes	5 x 10(5) CFU/ml	frankfurter sample	RAPTOR – uniform	Waveguide	Fluorescent sandwich assay	[31]
Listeria monocytogenes	5.4 x 10(7) CFU/ml	Hotdog samples	RAPTOR – uniform	Waveguide	Fluorescent sandwich assay	[30]
Listeria monocytogenes	4.3x10(3) CFU/ml	Buffer	Uniform	MM polystyrene	Fluorescent sandwich assay	[29]

Abbreviations: BT = Biconical Taper, TT = Tapered Tip, CTT = Combination Taper Tip, SM = Single Mode, MM = Multimode, ND = Not Described, * = the change in dB per hour per number of cells at inoculation, NA = Not Applicable.

Table 2. TFOBS For Biochemical Measurements

Target Analyte	LOD	Matrix	Taper Geometry	Fiber Type	Detection Principle	References
NADH, NADPH Concentration	Min: 0.2 μM (NADH), 0.5 μM (NADPH)	buffer	BT	SM	Absorption	[18]
Chinese Hamster Ovary Cell Concentration	Min: 10^5 cells/ml	buffer	BT	SM	Absorption	[18]
Paraoxon	Sub ppm	buffer	TT	MM	Chemiluminescence	[62]
STAT3	ND	Buffer	Uniform	MM	Fluorescent sandwich assay	[63]

Abbreviations: BT = Biconical Taper, TT = Tapered Tip, CTT = Combination Taper Tip, SM = Single Mode, MM = Multimode, ND = Not Described.

Table 3. TFOBS For Clinical Measurements

Target Analyte	LOD	Matrix	Taper Geometry	Fiber Type	Detection Principle	References
Protein A	1 µg/mL	ND	NA (chip)	NA	Leaky wave (SPR)	[64]
BSA	10 fg/mL	Buffer	BT	SM	Intensity	[1]
BSA	7.4 ng/mL	buffer	Chip (NA)	Chip (NA)	SPR	[65]
BSA	2.5 µg/ml	Buffer	BT	ND (plastic clad silica)	Dye-protein complex absorption	[43]
Ovalbumin	2.5 µg/ml	Buffer	BT	ND (plastic clad silica)	Dye-protein complex absorption	[43]
Hemoglobin	2.5 µg/ml	Buffer	BT	ND (plastic clad silica)	Dye-protein complex absorption	[43]
IgG	20 fM	Buffer	TT	MM	Fluorescent competitive assay	[44]
IgG	75 pg/mL	Serum and jejunal fluids diluted with buffer	BBT	SM	Fluorescent sandwich assay	[45]
Protein C	0.1 µg/mL	Buffer	TT	ND	Fluorescent sandwich assay	[16]
Protein C	0.5 µg/mL	Plasma	TT	MM	Fluorescent sandwich assay	[46]
Protein C	0.5 µg/mL	Plasma	Uniform	MM	Fluorescent sandwich assay	[66]
Protein C	0.5 µg/mL	Plasma	Uniform	MM	Fluorescent sandwich assay	[66]
Protein S	0.5 µg/mL	Plasma	Uniform	MM	Fluorescent sandwich assay	[66]
Antithrombin III (ATIII)	30 µg/mL	Plasma	Uniform	MM	Fluorescent sandwich assay	[66]

Table 3. Continued

Target Analyte	LOD	Matrix	Taper Geometry	Fiber Type	Detection Principle	References
Plasminogen (PLG)	30 µg/mL	Plasma	Uniform	MM	Fluorescent sandwich assay	[66]
B-type natriuretic peptide (BNP)	0.1 ng/mL	Plasma	Uniform	MM	Fluorescent sandwich assay	[66]
cardiac troponin I (cTnI)	1 ng/mL	Plasma	Uniform	MM	Fluorescent sandwich assay	[66]
C-reactive protein (CRP)	1 µg/mL	Plasma	Uniform	MM	Fluorescent sandwich assay	[66]
Myoglobin (MG)	75 ng/mL	Plasma	Uniform	MM	Fluorescent sandwich assay	[66]
L. donovani Antibody Concentration	Min: 0.244 ng/ml	Serum	CTT	MM (plastic clad silica)	Fluorescent sandwich assay	[17]
Progesterone	ng/mL	Buffer	ND	ND	Fluorescent sandwich assay	[42]
Adriamycin	0.01 µg/mL	blood	Straight core tip	MM	Fluorescence quenching	[49]
Cytochrome c	ND	Cell	TT	MM	Fluorescent sandwich assay	[50]
Cytochrome c	2.5 µg/ml	buffer	BT	ND (plastic clad silica)	Dye-protein complex absorption	[43]
Yersinia pestis fraction 1	50 ng/mL	Buffer, serum, plasma, and whole blood	BT	MM	Fluorescent sandwich assay	[15]
cTnI	31 pM	plasma	TT	MM quartz	Nano gold particle enhanced fluorescence	[67]
BNP	26 pM	plasma	TT	MM quartz	Nano gold particle enhanced fluorescence	[67]

Target Analyte	LOD	Matrix	Taper Geometry	Fiber Type	Detection Principle	References
Intracellular Benzopyrene Tetrol	6.4 pM	cell	TT	ND	Autofluorescence	[51]
Benzo\c\phenanthridinium alkaoids	ND	buffer	Chip	NA	SPR	[68]
Fumonisin B$_1$	10 ng/ml	methanol/water-extracted corn	TT	MM (plastic clad silica)	Fluorescent sandwich assay	[69]
Myoglobin	2.9 ng/mL	buffer	tip	MM	SPR	[39]
Myoglobin	5 nmol/L	buffer	Uniform probe	MM	Fluorescent Energy Transfer	[40]
Thrombin	1 nM	Buffer	Spheres (NA)	NA	Fluorescent competitive assay	[48]
Thrombin	1nM	Buffer	Uniform	MM	Coagulation of fluorescently labeled fibrinogen to unlabelled fibrinogen bound to the surface of the fibre optic	[47]
RNA	pM	Buffer	TT	SM	Fluorescence	[56]
DNA	70 fM	Buffer	Uniform	MM	Fluorescence	[52]
interleukin-1 (IL-1), interleukin-6 (IL-6), and tumor necrosis factor-α (TNF-alpha)	1 ng/mL	Buffer and spiked cell culture medium (CCM)	ND	MM	Fiber-optic surface plasmon resonance (SPR)	[41]
DNA	5 nM	buffer	ND	MM	Fluorescence	[53]

Abbreviations: BT = Biconical Taper, TT = Tapered Tip, CTT = Combination Taper Tip, SM = Single Mode, MM = Multimode, ND = Not Described, NA = Not Applicable.

3.1. Pathogen Detection

Escherichia coli O157:H7 [2, 23-26], *Salmonella typhimurium* [27, 28], *Listeria monocytogenes* [29-31], and *Bacillus anthracis* [32] are some of the pathogens which have been detected using TFOBS. Most pathogen detection studies done to date used fluorescence TFOBS [25-32], but a few of them used intensity-based TFOBS [2, 23, 24].

Ferreira et al. developed an intensity-based evanescent sensor, to be used with a 840 nm light source, to detect *Escherichia coli O157:H7* growth [23]. This evanescent sensor was fabricated by chemically etching. Transmission is reduced due to light absorption by the bacteria, and the power loss is proportional to the intrinsic bulk absorption and scattering, which depends on the concentration of the bacteria. The sensitivity of this sensor was 0.016 dB / hour-N_o, where N_o is initial cell concentration and ranges from 10 to 800. Similarly, Maraldo et al. used TFOBS to detect *Escherichia coli JM 101* growth on poly-L-lysine [24]. *E.coli JM 101* expressing green fluorescent protein was immobilized on the poly-L-lysine coated fibers, and growth was monitored by light transmission at 480 nm. The transmission decreased exponentially with cell growth on the tapered surface. In a follow up study by Rijal et al., *Escherichia coli O157:H7 (EC)* was covalently bonded to the surface of a TFOBS via an antibody, and concentrations as low as 70 cells/mL was detected by changes in intensity at 470 nm [2]. Detection of *EC* in real samples is of great interest and was investigated by DeMarco et al. [25]. *EC* in seeded ground beef samples was prepared and detected by a sandwich immunoassay using cyanine 5 dye-labeled polyclonal anti-*E. coli O157:H7*. Light was launched at 635 nm and the fluorescence was emitted at 670 to 710 nm. Responses were obtained within 20 minutes, and *E. coli O157:H7* at 3 to 30 CFU/mL were detected. A similar study was recently conducted by Geng et al., where a sandwich immunoassay was used with FOBS to detect *EC* in ground beef [26]. Light was launched at 635 nm and the fluorescence was emitted at 670 to 710 nm. The sensor detected 10(3) CFU/ml of pure cultured *EC* grown in culture broth. Artificially inoculated *EC* at concentration of 1 CFU/ml in ground beef was detected after 4 hours of enrichment.

Kramer et al. [27] studied the detection of *Salmonella typhimurium* in sprout rinse water using RAPTOR™, an evanescent fluorescence sensor developed by Research International, Monroe, Washington.. Alfalfa seeds contaminated with various concentrations of *Salmonella typhimurium* were sprouted, and the sprout water was measured by the instrument. *Salmonella typhimurium* was identified for seeds that were contaminated with 50 CFU/g. Zhou et al. [28] also used a sandwich immunoassay to detect *Salmonella*. Light was launched at 650 nm and the fluorescence was emitted at 680 nm. Tapered fiber tips with various geometries and treatments were studied and optimized, and *Salmonella* was detected at 10(4) CFU/mL.

An antibody-based sandwich fluorescence FOBS was developed by Geng et al. to detect *Listeria monocytogenes* [29]. Light was launched at 635 nm and the fluorescence emission was in the range of 670 to 710 nm. The sensor was specific, as shown by the significantly lower signals caused by other *Listeria* species or microorganisms. The LOD was 4.3x10(3) CFU/ml for a pure culture of *L. monocytogenes*. In less than 24 h, *L. monocytogenes* in hot dog or bologna was detected at 10 to 1,000 CFU/g after enrichment. Recently, Kim et al. also detected *L. monocytogenes* using the RAPTOR™ sensor [30]. This method achieved a LOD of 5.4 x 10(7) CFU/ml. *L. monocytogenes* was detected in phosphate buffered saline (PBS) by Nanduri et al. using RAPTOR™ to evaluate the effect of flow on antibody immobilization [31]. Light was launched at 635 nm and the fluorescence was emitted at around 670 nm. It was found that both the static and the flow through mode method had a LOD of 1 x 10(3) CFU/ml. However, the effective disassociation constant and the binding valences for static modes were higher than for flow through method of antibody immobilization. The flow through mode was chosen to test real samples, and the LOD was 5 x 10(5) CFU/ml.

Bacillus anthracis, is a serious threat to national security. Tims et al. addressed the need to detect *Bacillus anthracis,* and achieved detection at a concentration of 3.2 x 10(5) spores/mg in spiked powders in less than 1 hour [32]. The method used was based on fluorescent sandwich assay and a polystyrene tapered fiber. The excitation wavelength was 635 nm.

3.2. Toxin Measurement

TFOBS have been used to detect toxins such as enterotoxins [33-36], ricin [37], and endotoxins [38]. Fluorescence was used for all the toxins measurements which are discussed here.

Staphylococcal enterotoxins are a major cause of food poisoning. Tempelman et al. quantified *Staphyloccccocal* enterotoxin B (SEB) in a fluorescent sandwich immunoassay on a fiber optic biosensor [33]. A 635 nm diode laser was used to excite the labeled antibody. The fluorescence level was measured and gave a detection limit of 0.5 ng/mL. Shriver-Lake et al. used an array biosensor to detect SEB at a LOD of 0.5 ng/mL in buffer and six different types of food samples [34]. *Staphylococcus aureus* is the only species which produces protein A and was detected by Chang et al. using a fluorescent sandwich FOBS at a LOD of 1 ng/mL [35]. Excitation of this sensor was at 488 nm. Similar to SEB, *Clostridium botulinum* toxin A was detected by a fluorescent sandwich FOBS at 5 ng/mL [36]. A light source at 514 nm was used in this case.

Narang et al. reported a sandwich fluorescent TFOBS ricin detection in buffer and in river water [37]. The light source was 635 nm. Antibody to ricin was immobilized onto tapered fiber surface using silanization and avidin-biotin linkage. The avidin-biotin method had a higher sensitivity and wider linear dynamic range. The response of the avidin-biotin sensor was linear in the range of 100 pg/mL to 250 ng/mL. The LOD for ricin in buffer solution was 100 pg/mL, and in river water it is 1 ng/ml. At concentrations greater than 50 ng/ml, there was a strong interaction between ricin and avidin due to the lectin activity of ricin. This interaction was reduced for fibers coated with neutravidin or with the addition of galactose.

James et al. developed a method to detect lipopolysaccharide (LPS) endotoxin, which is the most powerful immune stimulant and causes sepsis [38]. LPS from *E. coli* was detected at a LOD of 10 ng/mL using fluorescent FOBS based on the competitive assay. Polymyxin B was used as a recognition molecule and was covalently immobilized onto the surface of the probe. Fluorescent labeled LPS was introduced to the fiber and attached to the Polymyxin B. Unlabeled LPS was then introduced and competed with the labeled LPS for the binding sites on the Polymyxin B. As LPS concentration increases, fluorescence decreases.

3.3. Clinical Measurements

Most clinical measurements done with TFOBS used proteins as analytes. Notable examples include cardiac markers [39, 40], cytokines [41], and hormones [42]. Investigators have detected model proteins using TFOBS in order to characterize TFOBS' potential. Preejith et al. detected model proteins using fiber optic evanescent wave spectroscopy [43]. They immobilized Comassie Blue on a multimode fiber surface using a porous glass coating. Comassie Blue normally absorbs at 467 nm, but it forms a dye-protein complex with the protein when exposed to an acidic environment, and such a complex absorbs at 590 nm. The protein concentration is inversely proportional to the output power at 590 nm, because increase in protein concentrations causes the evanescent absorption to increase. Calibration curves were obtained for BSA, hemoglobin, ovalbumin, and cytochrome c in the range of 0 to 20 µg/mL. In our laboratory, BSA was recently detected at 10 fg/mL in stagnant condition using intensity-based TFOBS [1].

Tromberg et al. detected antibody to IgG at 20 fM on a fluorescent FOBS tip using a competitive assay [44]. Light was launched at 488 nm and the fluorescence was emitted at 520 nm. Rabbit IgG was immobilized on the fiber tip, and exposed to fluorescein isothiocyanate (FITC) labeled and unlabeled anti-IgG. The response was inversely proportional to the amount of unlabeled anti-IgG, because the unlabeled anti-IgG displaced the labeled one. Hale et al. developed a fluorescent optical fiber loop sensor to detect antibody to IgG [45]. The sensor was used with a two-step sandwich assay. IgG was labeled with the fluorescent dyes fluorescein isothiocyanate or tetramethyl rhodamine. Antibody to IgG was detected at 75 pg/mL with this method.

Deficiency in Protein C (PC), if left untreated, may result in thrombotic complications, and, thus presents an important clinical challenge. Spiker et al. detected PC at 0.1 µg/mL in buffer using a sandwich fluorescent fiber optic sensor [16]. Real-time detection of PC in plasma is an important challenge in the clinical setting. Convective flow plays a vital role in the transport of PC in a viscous medium such as plasma. Tang et al. who examined PC detection in plasma with fluorescent sandwich FOBS and obtained a detection limit of 0.5 µg/mL [46].

Cardiac markers myoglobin (MG) and cardiac tropinin I (cTnI) can be measured to predict the occurrence of myocardial infarction, because they are released from cardiac muscles when they are damaged. A fiber-optic SPR sensor was developed by Masson et al. to detect MG and cTnI at 3 ng /mL [39]. A direct fluorescence FOBS was also used to detect myoglobin at 5nM [40]. An excitation wavelength of 425 nm was used to excite the Cascade Blue-labeled antibody, which was entrapped in the sensing element and fluoresces at 425 nm. Fluorescence quenching occurred when myoglobin attaches to the labeled antibody. Recently, Tang et al. developed a fiber-optic multi-analyte system which simultaneously quantifies two groups of multi-biomarkers related to cardiovascular diseases (CVD): anticoagulants (protein C, protein S, antithrombin III, and plasminogen) for deficiency diagnosis; and cardiac markers (B-type natriuretic peptide, cardiac troponin I, myoglobin, and C-reactive protein) for coronary heart disease diagnosis.

Garden et al. detected thrombin at 1 nM using fluorescent FOBS [47]. Excitation was at 495 nm and emission was at 520 nm. Unlabeled fibrinogen was first attached to the FOBS surface. Then, coagulation of solution phase fluorescently labeled fibrinogen to unlabelled fibrinogen bound to the surface was observed. Lee et al. detected thrombin at 1 nM using a fluorescent FOBS immobilized with an antithrombin DNA aptamer receptor [48]. The aptamer was immobilized on the surface of silica microspheres, which were distributed in microwells on the distal tip of an imaging fiber that was coupled to a modified epifluorescence microscope system. Another set of microspheres was prepared with a different oligonucleotide to measure the non specific binding. The distal end of the imaging fiber was incubated with fluorescein-labeled thrombin (F-thrombin), and the non-labeled thrombin was detected using the competitive method.

Progesterone was found to have evidence of carcinogenicity based on animal studies. Progesterone can be found in various surface waters commonly used for drinking water. In a study by Tschmelak et al., a fluorescence FOBS was immobilized with a labeled-antibody and used successfully to detect progesterone at concentrations lower than ng/L [42].

A fluorescent tip FOBS was used to measure adriamycin (ADM) at 10 ng/mL in vivo in a blood vessel [49]. A polymeric fluorescent D-70 membrane with pore sizes of 1-2 μm was immobilized on the fiber tip. Fluorescence was quenched by ADM present in the blood and the fluorescence signal was measured by a photomultiplier tube (PMT) at a wavelength of 530 nm.

The protein cytochrome c is involved in apoptosis and was detected by a sandwich fluorescent nanobiosensor fabricated by Song et al. [50]. δ-Aminolevulinic acid (5-ALA), a photodynamic therapy (PDT) drug, was activated by a He-Ne laser at 632.8 nm to induce apoptosis in MCF-7 human breast carcinoma cells. When mitochondria are damaged by PDT, cytochrome c is released into the cytoplasm; therefore cytochrome c concentration is an indication of apoptosis. Results indicate that 5-ALA PDT-treated cells had a much higher fluorescence signal, pointing to high cytochrome c concentrations in the treated cells.

Yersinia pestis is an etiologic agent of plague. A sandwich fluorescent FOBS devised by Cao et al. was used to detect *Yersinia pestis* Fraction 1 antigen at a limit of 5 ng/mL [15]. The light source was a 514 nm argon ion laser. This system detected $50 - 400$ ng/mL of protein in serum, and the results were in excellent agreement with ELISA results.

Nath et al. developed a fluorescent FOBS to detect *L. donovani* specific antibodies [17]. The sensor was made by de-cladding an optical fiber so that the evanescent wave propagated outside the tapered region. The sensor was used with a 488 nm light source. Cell surface protein of *L. donovani* was immobilized covalently on the sensing region. Then, the sensor was incubated with patient serum for 10 minutes, followed by incubation with goat anti-human IgG tagged with FTIC, which excites at 525 nm. The amount of *L. donovani* specific antibodies in the patient serum was proportional to the fluorescence. There were no false positive results from leprosy, tuberculosis, typhoid, and malaria serum.

Cullum et al. detected benzo[a]pyrene tetrol (BPT) at 6.4 ± 1.7 E pM in mammary carcinoma cells using a sandwich fluorescent fiber-optic nanosensor tip [51]. BPT is a metabolite of benzo[a]pyrene. Using a 325 nm light source, the authors were able to calibrate the sensor and obtain an unknown concentration by observing the level of fluorescence. This technique is useful for cancer screening since carcinogens bind to DNA and form substances such as BPT.

Three cytokines related to chronic wound healing are interleukin-1 (IL-1), interleukin-6 (IL-6), and tumor necrosis factor-α (TNF-alpha) [41]. A fiber-optic SPR sensor was modified with antibodies at the surface, and detected these proteins with LOD of 1 ng/mL in buffered saline solution and spiked cell culture medium (CCM).

3.4. DNA Hybridization

Kleinjung et al. detected DNA hybridization at 3.2 attomoles (70 fM) using a fluorescent multimode FOBS with 13-mer probe attached to the de-cladded core [52]. The complementary strands were labeled and detected when introduced to the sensor. This sensor was able to distinguish between matching sequences, single nucleotide mismatch, and mismatch caused by additional deviations.

Zeng et al. examined the interfacial hybridization kinetics of oligonucleotides immobilized onto silica using a fluorescent FOBS that was excited at 632 nm [53]. A dT20 DNA probe was used as recognition molecules, while target fluorescein-labeled non-complementary DNA (ncDNA) dT20 and fluorescein-labeled dA20 were detected. The target DNA concentrations were 5 nM to 0.1 μM. The response of the sensor fit the second order Langmuir model.

Molecular beacons (MB) are oligonucelotide probes that fluoresces upon hybridization with target DNA or RNA molecules [54]. Liu et al. immobilized MB on a fluorescent FOBS and determined the effects of ionic strength and target DNA concentration on hybridization kinetics. Using an excitation wavelength of 514 nm, they found the LOD was 1.1 nM of DNA. The sensor showed selectivity by distinguishing between 100 nM of ncDNA, 100 nM of one-base mismatch, and 100 nM of cDNA [54].

3.4.1. Pathogen Detection via DNA

A fluorescent FOBS was developed by Almadidy et al. to detect short sequences of oligonucleotides that identify *E. coli* microbial contamination [55]. DNA probes were first immobilized to silica surface via a silane reagent. Then, stepwise synthesis of oligonucleotides by the β-cyanoethyl-phosphoramidite protocol took place on the surface. The sensor was exposed to both complementary (cDNA) and non-complementary (ncDNA) 20-mers, as well as genomic DNA from *E.coli*. The cDNA and ncDNA were introduced at a concentration of about 1.7 nM, whereas genomic DNA was introduced at 1.7 pM to 170 pM. Fluorescent intercalating dye was used to detect hybridization. Quantities as low as 100 fM were detected using this method.

Pilevar et al. detected *Helicobacter pylori* total RNA using a fluorescent FOBS that had probes immobilized on its surface [56]. IRD-41 is a near-infrared fluorophore which is excited by 785 nm light. Real-time hybridization measurement of IRD 41-labeled oligonucleotide at various concentrations to the surface bound probes was performed. Complementary DNA at lower than nM concentration was detected. Sandwich assays were performed with *Helicobacter pylori* total RNA, and results showed that this sensor could detect *H. pylori* RNA in a sandwich assay at 25 pM.

4.0. Methods

4.1. Fabrication

Corguide fibers (Corning Glass Works, NY, attenuation at 1300 and 1500 nm of 0.36 and 0.26 dB km−1, respectively) with a core diameter of 8 μm and total diameter of 125 μm were used in all the fabrication methods described here. The fabrication methods commonly used in our laboratory are chemical etching, heat pulling by flame, and heat pulling by fusion splicer.

4.1.1. Chemical Etching

Chemical etching using hydrofluoric acid (HF) is one of the simplest ways to create tapers with a step change in radius. Acrylic (Plexiglas) was used to construct the etching reactor because HF does not attack most plastic materials. In order to monitor the etching, a spectrofluorometer was used to detect the transmission through the fiber as etching took place. This instrument has a compact 75 W Xenon arc lamp (Ushio Inc., Japan) coupled to a monochromator and a PMT (model R1527P in housing 710, PTI Inc.) coupled to a monochromator.

The plastic sheathing of the fiber was removed by immersing the fiber in acetone (Fisher Scientific) for 15–20 min followed by mechanical removal with a fiber optic stripper (NO-NIK). A fiber optic cleaver (NO-NIK) was used to make a clean-cut fiber tip so as to enhance the efficiency of light collection into the fiber.

HF (Fisher Scientific, Philadelphia) at a concentration of 49.5 wt.% was used. Two hundred microliter of HF was introduced into the reaction chamber. Once HF was injected, the transmission was monitored at 350 nm. When the diameter of the fiber was etched to a certain fraction of the initial diameter, the etching process was stopped by first removing the HF and then washing the chamber twice with 5 N NaOH as rapidly as possible. The fiber was then immersed in a 200 mL of 0.1 N NaOH bath for 60 min to stabilize the fiber. If this step was not carried out, any remaining HF would have continued to etch the fiber until it dissolved completely. It was found that the length of the etching time needed at room temperature was about 40 min.

4.1.2. Heat Pulling Using a Manual Propane Torch

Heat pulling using a micro-propane torch is another method of obtaining tapers. The apparatus for this method is illustrated in Figure 4. The polymeric sheathing of a 30-35 cm long fiber was removed similarly as in the chemical etching method. The fiber was then mounted on the apparatus with two paper clips of identical weights (2.8 g) on either ends to provide tension to the fiber. A micro-flame (Model 6000, Microflame, Inc., MN) was positioned such that the fiber was approximately one third distance from the top end of the visible end of the flame, and the flame was removed as soon as the paper clip touched the stop. The ends of the tapered fiber were cleaved using a fiber-optic cleaver (NO-NIK) to give clean cut ends. The fiber was placed in an optical fiber holder to be used in the experiments. The dimensions of a fiber were measured after taking micrographs of the taper using an IMT-2 optical microscope (Olympus, Japan) equipped with a video camera (Cohu Corp., Japan) linked to a computer. The dimensions were measured in the Scion Image software (Scion Corp., MD) after a calibration was performed according to the microscope objective used.

4.1.3. Heat Pulling Using a Fusion Splicer

The polymeric sheathing was removed over a distance of 5 cm at the center and both ends of the fiber. The fiber was cleaned with isopropanol and the ends were cut clean using a fiber cleaver (Ericsson EFC 11-4). The fiber was inserted into the programmable fusion splicer (Ericsson FSU975), where electric current was applied via a pair of electrodes for up to 60 seconds while the taper was pulled automatically. Various current levels (3-13 mA) and pull times (2-30 s) were used to produce fibers of varying taper diameters and lengths. A micrograph of the fiber was taken via a camera inside the fusion splicer, and the dimensions were measured in the Scion Image software (Scion Corp., MD).

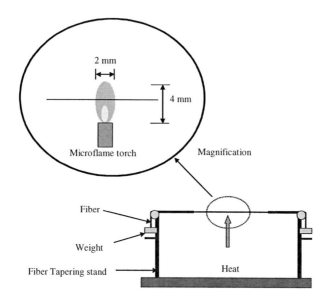

Figure 4. Some of the fibers discussed in this paper were fabricated using a simple fiber tapering device. Fiber without sheathing was mounted on a stand and two pieces of weights were attached to the ends to provide tension required for tapering. The fiber was then heated with a flame while carefully monitoring the diameter of the taper. When the desired diameters were reached, the two ends of the fibers were clipped and the tapered fiber were placed on an optical fiber holder to be used in the experiments. Adapted from [2].

4.2. Optical Characterization of Tapers

4.2.1. Preliminary Characterization Using Water

The preliminary characterization method used in our laboratory for determining the evanescent field strength is the comparison of the transmission in water to that in air. The reason for the choice of these two media was that they provide the most difference in refractive index that is expected to be present in the waist region for biological samples. If a taper exhibited little or no transmission change going from air to water, its transmission was not expected to change significantly at the presence of dilute analyte solutions.

4.2.2. Characterization in the Visible Range Using **E.coli JM101**

Characterization in the visible range was performed on fusion spliced and torch heat-drawn tapers using *E.coli JM101* (ECJ) as the analyte. Tapers that showed little or no transmission change in water compared to air, also showed no or low response to the presence of ECJ suspensions. Both symmetric and asymmetric tapers of small and large waist diameters had this behavior.

Several tapers exhibited a significant transmission difference (~50%) in water compared to air. These tapers also showed little or no change in the presence of the ECJ. Small RI changes due to the presence of ECJ suspension were not sufficient to produce the transmission changes, resulting in poor sensitivity. Any impact on the light through the fiber was already saturated from change due to water itself, such that the presence of ECJ suspension had little further impact. Other tapers that had this characteristic property, but were of smaller waist diameter showed weak sensitivity. Most of such tapers were symmetric tapers. On the other hand, the asymmetric tapers showed lower relative transmission through water, but allowed further modulation in transmission from the presence of ECJ.

We compared relative transmission at 470 nm for fusion splicer tapered fibers. In general, there were two types of responses. In the first type, the transmission increased or decreased monotonically, as ECJ concentration increased. In the second type, an initial increase for low cell concentration is followed by a decrease at higher cell concentration. There were tapers which also showed an increase in transmission at low ECJ concentrations, followed by decrease at intermediate concentrations and then an increase at 7 million cells/mL.

Heat drawn tapers have typically a much longer convergent, waist and divergent sections, each on the order of millimeters. Similar to the fusion spliced tapers, HD tapers showed two basic characteristics. In one case, tapers showed a decrease in transmission as concentration increased. In the other case, tapers showed a slight increase and then a decrease in transmission for higher concentrations. At low concentration, the RI of cellular suspension influenced transmission response, and caused the increase in transmission. At higher concentrations, the evanescent light absorption by the cells dominated the response. These results suggest that torch-drawn tapers have excellent potential as biosensors.

4.2.3. Characterization in the RI Range Using Glucose Solutions

In order to characterize the tapered fibers in the IR region, we measured transmission properties under various RI fluids in the tapered region, using an experimental setup similar to Figure 5 but in flow condition. The transmissions at 1310 and 1550 nm were monitored and recorded simultaneously using a spectrum analyzer and LabView program. Once the transmission stabilized in air, de-ionized (DI) water was flowed in at 0.5 mL/min. Glucose solutions of various concentrations were then flowed past the taper, with de-ionize (DI) water flowed in to rinse out the taper in between glucose solutions.

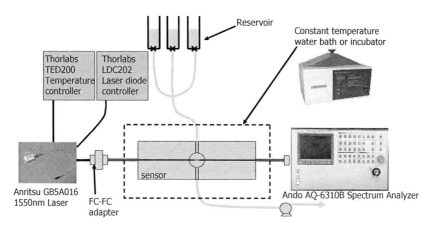

Figure 5. Experimental setup at 1550 nm in stagnant condition.

4.3. Antibody Immobilization

The most commonly used antibody immobilization method in our laboratory was adapted from Hermanson [57] with modification for the fiber surface and geometry. Prior to immobilization, the taper was cleaned with 1 M hydrochloric acid for 30 minutes, sulfuric acid for 10 minutes, and 1 M sodium hydroxide for 10 minutes. The sample holder and taper were rinsed several times with de-ionized water between cleaning steps. The cleaning procedure produced reactive hydroxyl groups on tapered surface. The surface was then silanylated with 3-aminopropyl-triethoxysilane (APTES; Sigma-Aldrich) in de-ionized water for 2-24 hours. The fiber was then dried overnight in a vacuum oven at 40°C, or in a regular oven at 75°C. The APTES reaction creates amine groups at the surface, which can further react with carboxylic groups in the antibody to form a peptide bond. The polyclonal antibody to BSA (anti-BSA; Sigma Catalog # B1520) contains carboxyl groups which were activated using 1-ethyl-3-(3-dimethylaminopropyl)-carbodiimide (EDC; Sigma-Aldrich) and stabilized by sulfo-N-hydroxysuccinimide (Sigma-Aldrich). EDC converts carboxylic groups into reactive unstable intermediates which are susceptible to hydrolysis. However,

Sulfo-NHS replaces the EDC, resulting in a more stable reactive intermediate which catalyzes reaction with amine groups. To prepare the antibody, 0.4 mg of EDC and 1.1 mg of sulfo-NHS was added to each mL of antibody solution and the reaction was left on for 30 minutes at room temperature. Then, 1.4 μL of 2-mercaptoethanol was added to quench the EDC. This intermediate was added to the silanylated tapered fiber surface and covalent bonding was carried out at room temperature for 2 hours, in stagnant condition. At the end of antibody immobilization, Hydroxylamine was added to regenerate the carboxylic groups of the antibody. Transmission through the fiber was recorded during antibody immobilization and is shown in Figure 6.

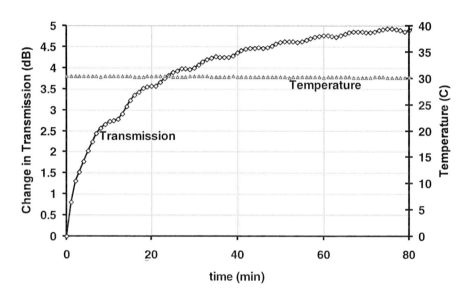

Figure 6. Transmission change vs. time for antibody immobilization at 1550 nm. Temperature was held constant at 30 °C ± 0.5 °C as indicated. Adapted from [1].

Alternatively, the antibody can be activated via carbohydrate groups. For this protocol, 1 mg/ml of antibody was dissolved in PBS and protected from light. Then, 100 μL of 0.1 M NaIO4 solution was added to antibody and allowed to react for 30 minutes. The silanized fibers were exposed to the solution for 2 hours. Then, 10 μL of NaCNBH3 was added for 30 minutes to reduce the Schiff Base to a second amine.

Another possible method of functionalization which is currently under investigation is the use of Protein G with gold. In this method, the taper was first coated with a 1:1000/v:v Polyurethane/Toluene mixture and dried overnight. The taper was then coated with 10 to 100 nm of gold using Denton Vacuum Desk IV® system. After gold coating, the taper was enclosed in the fiber holder by epoxy. During the first step of immobilization, Protein G was flowed into the sample chamber and left there in stagnant condition for 90 minutes. The sample chamber was then rinsed thoroughly with PBS, and antibody was flowed in and left there for 90 minutes. Then, the chamber was rinsed thoroughly prior to using it in a detection experiment.

4.4. Sample Preparation

All biological samples were prepared as per the instructions of the manufacturer using solutions of 0.1% Sodium Azide in PBS as the solvent. Usually a bulk solution of antibody is made and then aliquots of 4 mL are dispensed into sterilized scintillation vials. The vials are then stored at -30 C freezer until use. The antibody vials were for single use and were disposed at the end of the experiment. As for the analytes such as BSA and E.coli, they were prepared in bulk in sterile centrifuge tubes each holding a maximum of 50 mL. Each tube contained one concentration of analyte, and they were all stored at $4^{\circ}C$. The tubes were placed back refrigerated at $4^{\circ}C$ after each use.

4.5. Detection

4.5.1. E.coli O157:H7 in Stagnant Condition

E.coli O157:H7 (EC) was detected using a wavelength of 470 nm in stagnant conditions. Tapers were fabricated using heat pulling by torch or fusion splicer. The surfaces of the tapers were functionalized with antibody to *E.coli O157:H7* using APTES and carboxylic linkage. The taper was exposed to various concentrations of pathogen, and showed transmission changes as the antigen attached.

An EC stock solution ($7x10^9$ cells/mL) was prepared as per the vendor's (KPL) rehydration protocol in 10 mM PBS at pH 7.4. Lower concentrations ($7x10^7$ cells/mL, $7x10^5$ cells/mL, $7x10^3$ cells/mL, and 70 cells/mL) were prepared in PBS (pH 7.4) by serial dilution. 150 μL of each sample was injected into the sample chamber. After EC attachment, the sample was removed and the chamber was loaded with either HCl/PBS buffer at pH of 2.3 or Glycine-HCl/ethylene glycol (1:1 v/v) buffer at pH 1.7.

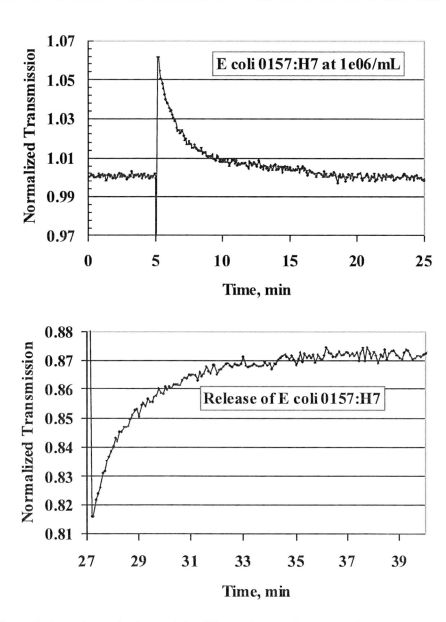

Figure 7. Detection and release of 1 million cells/mL of EC on antibody immobilized tapered fiber. EC detection and release experiment were performed on a 8.8 μm diameter TFOBS. After attachment (top panel), release buffer (glycine-HCl/ethylene glycol buffer, pH 1.7) was injected into the chamber to release EC (bottom panel).

The response due to attachment and release of EC cells are shown in Figure 6. Immediately upon addition of the 1E6 cells/mL of EC sample, there was a rapid increase in transmission due to the RI change of the medium. Subsequently, a gradual and exponential decrease in transmission occurred due to EC attachment. Cells change the RI surrounding the fiber and absorb light from the evanescent field. When the attachment reached equilibrium, no further light is absorbed and the transmission remained constant.

The antigen attached to the sensor may be released by altering the pH as the antibody-antigen binding is pH-dependent. The response due to release was equal in magnitude and opposite in direction, as shown in Figure 7. This change occurred because cells released into the bulk are too far away from the taper surface to influence light transmission. When cells released reached equilibrium, the transmission reached a constant value.

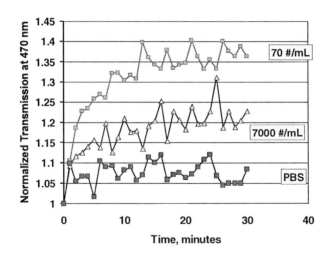

Figure 8. The response at 470 nm due to different concentrations of EC cells. Adapted from [2].

Intuitively, one would imagine that the transmission change would be directly proportional to the concentration. However, results show that the magnitude of the change is inversely proportional to the pathogen concentration, as shown in Figure 8. In addition, the response for this experiment was an increase in transmission, contrary to the experiment shown in Figure 7. The cause of this is not entirely clear, but we believe that it is due to the combined effects of evanescent absorption and scattering of the evanescent light. As cells cover the taper surface, the evanescent light is absorbed by the cells in proportion to the surface coverage. On the other hand, the RI is increased due to cell attachment. If the sample were homogeneous, increase in refractive index tends to increase transmission through the core due to reduction in the evanescent field. Hence, cell attachment result in transmission increase or decrease.

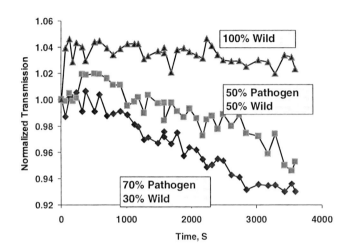

Figure 9. Effect of pathogenic and non-pathogenic EC mixture. Experiments were performed on a 9.5 µm diameter TFOBS. When 0% pathogen (100% wild strain JM101) was injected around the taper, there was no significant transmission change through the taper. When a solution containing an EC and the wild strain is added to the solution, EC bind to the antibody thus resulting in a decrease in transmission through the fiber. As the concentration of EC is increased to 50% and 70%, there is a greater binding of pathogen to the antibody on the surface and thus greater change in transmission occurs. Adapted from [2].

In order to evaluate specificity, the response to non-pathogenic *E. coli* was measured. Stock solution containing EC was mixed with a wild strain of *E. coli* (JM101) in volumetric proportions of 0%, 50% and 70%. The total bacterial count was 7×10^7 cells/mL. The detection experiments were carried out in the same manner as with pure EC. The sensor showed good selectivity to the pathogenic antigen as shown in Figure 9.

It is useful to obtain the kinetics EC attachment on antibody-immobilized surfaces. The immobilization and detection responses show exponential behavior, similar to the adsorption process often referred to as Langmuir kinetics. The Langmuir kinetics model can be expressed as [58]:

$$\theta = 1 - e^{-k_{obs}t} \tag{11}$$

where $\theta \ (0 \le \theta \le 1)$ is the fractional coverage of the reactive sites at time t. The parameter, k_{obs}, is the observed binding rate constant, which depends on the bulk concentration of the reactant. We hypothesize that the transmission is indicative of attachment, and express the Langmuir model as follows:

$$\left(\Delta I\right) = \left(\Delta I_\infty\right)\left(1 - e^{-kC_b t}\right) \tag{12}$$

where $\left(\Delta I\right)$ is the transmission change at time, t, $\left(\Delta I_\infty\right)$ is the steady state transmission change, and C_b is the bulk concentration. Taking the natural log on both sides of Eq. (12) we obtain:

$$\ln\left(\frac{\left(\Delta I_\infty\right) - \left(\Delta I\right)}{\left(\Delta I_\infty\right)}\right) = -kC_b t \tag{13}$$

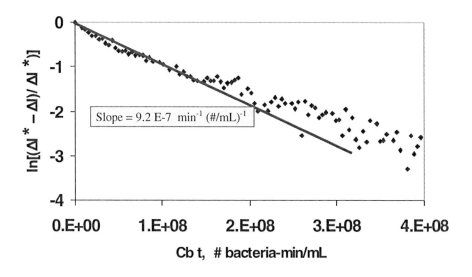

Figure 10. Calculation of rate of attachment (slope k) for EC.

The above suggests that the characteristic rate constant k during initial time can be determined from a plot of the left hand side versus C_b*t in Eq. (13). Figure 10 is an example of a graph displaying Eq. (13). The kinetic constant (k) was found to be in the range of 4×10^{-9} min^{-1} (pathogen/mL)$^{-1}$ to 7×10^{-9} min^{-1} (pathogen/mL)$^{-1}$.

4.5.2. BSA in Stagnant Condition

Although we were able to detect E.coli O157:H7 at 470 nm, the sensitivity of the sensors was limited due to the diameter in relation to the wavelength. Because the fibers are very fragile, we are unable to fabricated tapers that are less than 5 μm in diameter. However, at 470 nm the penetration of the evanescent field is limited. Also, cells are relatively large compared to the evanescent field generated at 470 nm. On the other hand, according to Eq. (1), there are reasons to believe that the evanescent field would be larger at a longer wavelength. Therefore, detection of BSA was detected similarly to EC but performed mostly using near-IR wavelengths. We first reported the use of a 1550 nm laser with TFOBS to monitor the real-time attachment of BSA to the antibody-immobilized surface [1]. While cuvette measurements established that BSA was non-absorbing at 1550 nm, antibody-immobilized TFOBS showed transmission changes at bulk concentrations of 10 fg/mL of BSA. The experimental setup for near-IR detection is shown in Figure 5.

Solutions of BSA from 10 fg/mL to 1mg/mL were prepared. After antibody was immobilized, it was rinsed with PBS, and 200 μL of BSA was injected into the sample chamber. Only one concentration was used in each experiment of attachment and release. After attachment, the BSA was removed and the sensor was rinsed with PBS. Then, PBS adjusted to a pH of 2 by H_2SO_4 was added to release the BSA. The acidic PBS weakens the binding of BSA to the antibody because it changes the conformation of the protein. After this the fiber surface was regenerated with the cleaning sequence, followed by modification by APTES.

When BSA was injected into the sample holder, transmission decreased due to change in surface refractive index caused by the presence of BSA. When BSA was replaced by low pH PBS, transmission increased back almost to the starting value. Similar experiments were performed with many tapers using different concentrations, and we conclude that experimental results are reproducible with the different fibers. The results of the attachment and release of 10 pg/mL of BSA are shown in Figure 11.

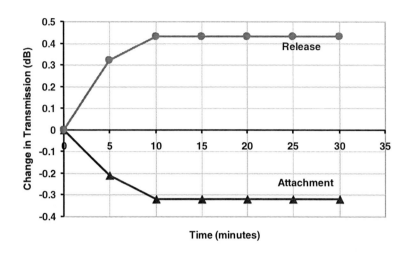

Figure 11. BSA attachment and release of 10 pg/mL sample. The attachment response was obtained when the tapered region was first exposed to 10 pg/mL of BSA. Data was collected for 30 minutes, rinsed with PBS, and then the tapered region was exposed to pH2 PBS for BSA release. The transmission changes due to attachment and release are in opposite direction and have approximately the same magnitude. Adapted from [1].

Multiple step attachment experiments was performed on three TFOBS with up to five different solutions of BSA, with concentration ranging from 100 fg/mL to 10 ng/mL. The experiments were initially performed with a starting concentration of 100 fg/mL. In the last experiment, shown in Figure 12, the initial concentration was set at 10 fg/mL. The BSA solutions were added in order from the lowest to the highest concentration, with removal of each sample after collection of data for up to 40 minutes. The transmission decreased as a function of time as BSA attached to the antibody. The steady state transmission for each concentration also decreased.

Like the EC results, transmission change is not linearly proportional to concentration. We believe that the reason for this is that at low concentrations, the surface of the fiber is not saturated with the antigen BSA. An estimate of the antibody/antigen surface coverage can be made with a few simplifying assumptions, and it was suggested that the concentration required for saturation is less than 4 ng/mL.

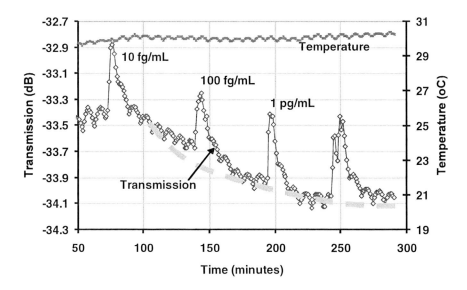

Figure 12. Semi-batch staircase experiment showing attachment of BSA from 10 fg/mL to 10 pg/mL. Temperature was maintained at 30 °C ± 0.5 °C using an incubator. The BSA solutions were added sequentially from lowest to highest upon removal of the previous solution. The purple line with peaks represents transmission through the fiber. The peaks correspond to time instants when the samples were introduced. The dotted line at the bottom represents the trend exhibited by the steady state transmission with respect to time. Adapted from [1].

Transmission changes are caused by the evanescent field interaction with the surface layer of antigen. Once the concentration approaches ng/mL levels, the surface is saturated with BSA. Additional BSA molecules would attach on top of the surface layer. However, the evanescent field magnitude decays away from the surface. Therefore the effect of BSA on top of the first layer results in much smaller changes. In addition, the condition for immobilization varies from one experiment to another. It is possible that nonlinearity was observed because at the lowest concentration, the bulk refractive index is approximately the same as that of PBS, and the BSA molecules on the fiber surface act as isolated points of high refractive index. When the concentration increases to saturation point, the fiber surface is covered with a layer of BSA which has a higher refractive index than PBS.

5.0. CONCLUSION

It is seen that TFOBS have several advantages in in terms of detection, including sensitivity, selectivity, ease of use, affordability, ability for remote sensing, and small sample volumes. They have been used for many applications such as pathogen detection, medical diagnostics based on protein or cell concentration, and detection of DNA hybridization.

As far as the sensor physics is concerned, intensity-based sensors have been used to a limited extent in cell detection. On the other hand, fluorescence based TFOBS are widely used for protein and DNA detection because amplification is a convenient tool, and often necessary to achieve low LODs. In addition, SPR is commonly used for protein characterization and has also been used for the detection of DNA hybridization. The ng/mL LOD of SPR makes it suitable for many medical applications. While fluorescence is very selective, its LOD is higher than SPR's. In addition, fluorescence requires multiple steps for the preparation of the sensor or the sample.

In terms of target analytes, one possible area of growth is the use of SPR or intensity-based TFOBS protein and DNA detection. Another application may be drug screening using TFOBS. Because of recent concerns of homeland security, there will likely be a significant push for research in bio-threat detection. Pathogen detection also remains important in maintaining a safe environment and food supply. Clinical applications of TFOBS will likely be important as medical professionals seek convenient methods to diagnose diseases.

TFOBS have been used as intensity-based sensors in our laboratory. We have used three methods of fabrication: step-etching using hydrofluoric acid, heat pulling by flame, and heat pulling by fusion splicer. The sensing ability of TFOBS was characterized by measuring the transmission in water, *E.coli JM101* solutions, and glucose solutions. TFOBS were functionalized with antibodies using covalent bonding or surface coating with gold and Protein G.

TFOBS was used in our laboratory to measure *E.coli O157:H7* in stagnant condition. One surprising finding was that concentration had an inverse effect on the transmission. TFOBS was shown to be selective to the pathogens. BSA was detected at 10 fg/mL in stagnant condition at 1550 nm. Transmission data was fitted to the Langmuir absorption model to determine the attachment rate.

As TFOBS evolves, new efforts will be focused on enhancing the sensitivity and selectivity. Improved surface chemical modification and stability of the recognition molecule can increase the sensitivity and robustness of TFOBS, especially for intensity-based TFOBS because it is the most sensitive when molecules are bound to its surface. As was shown in this chapter, there is a solid foundation of work to support the use of TFOBS and a wide variety of applications. Given its promising advantages, it is likely that TFOBS will remain a popular choice for detection in the future.

Acknowledgments

This work was supported through the National Science Foundation Grant # CBET-0329793, "Ultra Sensitive Continuous Tapered Fiber Biosensors for Pathogens and Bioterrorism Agents".

REFERENCES

[1]   Leung, A., Shankar, P.M., Mutharasan, R. (2006) Real-time monitoring of bovine serum albumin at femtogram/mL levels on antibodyimmobilized tapered fibers *Sens. Actuators B: Chem.* in press, doi:10.1016/j.snb. 2007. 03.010 .

[2]   Rijal, K., Leung, A., Shankar, P.M., Mutharasan, R. (2005) Detection of vathoizen Escherichia coli O157 : H7 AT 70 cells/mL using antibody-immobilized biconical tapered fiber sensors. *Biosensors & Bioelectronics* 21: 871-880.

[3]   Leung, A., Rijal, K.S., Shankar, P.M., Mutharasan, R. (2006) Effects of geometry on transmission and sensing potential of tapered fiber sensors. *Biosensors & Bioelectronics* 21: 2202-2209.

[4]   Arregui, F.J., Matias, I.R., Lopez-Amo, M. (2000) Optical fiber strain gauge based on a tapered single-mode fiber. *Sensors and Actuators a-Physical* 79: 90-96.

[5]   Datta, P., Matias, I., Aramburu, C., Bakas, A., LopezAmo, M., Oton, J.M. (1996) Tapered optical-fiber temperature sensor. *Microwave and Optical Technology Letters* 11: 93-95.

[6]   Bariain, C., Matias, I.R., Arregui, F.J., Lopez-Amo, M. (2000) Optical fiber humidity sensor based on a tapered fiber coated with agarose gel. *Sensors and Actuators B-Chemical* 69: 127-131.

[7]   Diaz-Herrera, N., Navarrete, M.C., Esteban, O., Gonzalez-Cano, A. (2004) A fibre-optic temperature sensor based on the deposition of a thermochromic material on an adiabatic taper. *Measurement Science & Technology* 15: 353-358.

[8] Bariain, C., Matias, I.R., Arregui, F.J., Lopez-Amo, M. (2000) Tapered optical-fiber-based pressure sensor. *Optical Engineering* 39: 2241-2247.
[9] Villatoro, J., Luna-Moreno, D., Monzon-Hernandez, D. (2005) Optical fiber hydrogen sensor for concentrations below the lower explosive limit. *Sensors and Actuators B-Chemical* 110: 23-27.
[10] Villatoro, J., Diez, A., Cruz, J.L., Andres, M.V. (2003) In-line highly sensitive hydrogen sensor based on palladium-coated single-mode tapered fibers. *Ieee Sensors Journal* 3: 533-537.
[11] Bariain, C., Matias, I.R., Romeo, I., Garrido, J., Laguna, M. (2000) Detection of volatile organic compound vapors by using a vapochromic material on a tapered optical fiber. *Applied Physics Letters* 77: 2274-2276.
[12] Kim, Y.C., Peng, W., Banerji, S., Booksh, K.S. (2005) Tapered fiber optic surface plasmon resonance sensor for analyses of vapor and liquid phases. *Optics Letters* 30: 2218-2220.
[13] McCue, R.P., Walsh, J.E., Walsh, F., Regan, F. (2006) Modular fibre optic sensor for the detection of hydrocarbons in water. *Sensors and Actuators B-Chemical* 114: 438-444.
[14] Zalvidea, D., Diez, A., Cruz, J.L., Andres, M.V. (2006) Hydrogen sensor based on a palladium-coated fibre-taper with improved time-response. *Sensors and Actuators B-Chemical* 114: 268-274.
[15] Cao, L.K., Anderson, G.P., Ligler, F.S., Ezzell, J. (1995) Detection of Yersinia-Pestis Fraction-1 Antigen with a Fiber Optic Biosensor. *Journal of Clinical Microbiology* 33: 336-341.
[16] Spiker, J.O., Kang, K.A. (1999) Preliminary study of real-time fiber optic based protein C biosensor. *Biotechnology and Bioengineering* 66: 158-163.
[17] Nath, N., Jain, S.R., Anand, S. (1997) Evanescent wave fibre optic sensor for detection of L-donovani specific antibodies in sera of kala azar patients. *Biosensors & Bioelectronics* 12: 491-498.
[18] Haddock, H.S., Shankar, P.M., Mutharasan, R. (2003) Evanescent sensing of biomolecules and cells. *Sensors and Actuators B-Chemical* 88: 67-74.
[19] Ko, S.H., Grant, S.A. (2006) A novel FRET-based optical fiber biosensor for rapid detection of Salmonella typhimurium. *Biosensors & Bioelectronics* 21: 1283-1290.
[20] Bobb, L.C., Shankar, P.M., Krumboltz, H.D. (1990) Bending Effects in Biconically Tapered Single-Mode Fibers. *Journal of Lightwave Technology* 8: 1084-1090.
[21] Okamoto, K. (2000): In *Fundamentals of Optical Waveguides*, Academic Press, pp. 323-330.

[22] Moar, P.N., Huntington, S.T., Katsifolis, J., Cahill, L.W., Roberts, A., Nugent, K.A. (1999) Fabrication, modeling, and direct evanescent field measurement of tapered optical fiber sensors. *Journal of Applied Physics* 85: 3395-3398.

[23] Ferreira, A.P., Werneck, M.M., Ribeiro, R.M. (2001) Development of an evanescent-field fibre optic sensor for Escherichia coli O157 : H7. *Biosensors & Bioelectronics* 16: 399-408.

[24] Maraldo, D., Shankar, P.M., Mutharasan, R. (2006) Measuring bacterial growth by tapered fiber and changes in evanescent field. *Biosensors & Bioelectronics* 21:1339-1344.

[25] DeMarco, D.R., Saaski, E.W., McCrae, D.A., Lim, D.V. (1999) Rapid detection of Escherichia coli O157: H7 in ground beef using a fiber-optic biosensor. *Journal of Food Protection* 62: 711-716.

[26] Geng, T., Uknalis, J., Tu, S.I., Bhunia, A.K. (2006) Fiber-optic biosensor employing Alexa-Fluor conjugated antibody for detection of Escherichia coli O157 : H7 from ground beef in four hours. *Sensors* 6: 796-807.

[27] Kramer, M.F., Lim, D.V. (2004) A rapid and automated fiber optic-based biosensor assay for the detection of salmonella in spent irrigation water used in the sprouting of sprout seeds. *Journal of Food Protection* 67: 46-52.

[28] Zhou, C.H., Pivarnik, P., Auger, S., Rand, A., Letcher, S. (1997) A compact fiber-optic immunosensor for Salmonella based on evanescent wave excitation. *Sensors and Actuators B-Chemical* 42: 169-175.

[29] Geng, T., Morgan, M.T., Bhunia, A.K. (2004) Detection of low levels of Listeria monocytogenes cells by using a fiber-optic immunosensor. *Applied and Environmental Microbiology* 70: 6138-6146.

[30] Kim, G., Morgan, M.T., Ess, D., Hahm, B.K., Kothapalli, A., Valadez, A., Bhunia, A. (2006) Detection of Listeria monocytogenes using an automated fiber-optic biosensor: RAPTOR. *Advanced Nondestructuve Evaluation I, Pts 1 and 2, Proceedings* 321-323: 1168-1171.

[31] Nanduri, V., Kim, G., Morgan, M.T., Ess, D., Hahm, B.K., Kothapalli, A., Valadez, A., Geng, T., Bhunia, A.K. (2006) Antibody immobilization on waveguides using a flow-through system shows improved Listeria monocytogenes detection in an automated fiber optic biosensor: RAPTOR (TM). *Sensors* 6: 808-822.

[32] Tims, T.B., Lim, D.V. (2004) Rapid detection of Bacillus anthracis spores directly from powders with an evanescent wave fiber-optic biosensor. *Journal of Microbiological Methods* 59: 127-130.

[33] Tempelman, L.A., King, K.D., Anderson, G.P., Ligler, F.S. (1996) Quantitating staphylococcal enterotoxin B in diverse media using a portable fiber-optic biosensor. *Analytical Biochemistry* 233: 50-57.

[34] Shriver-Lake, L.C., Shubin, Y.S., Ligler, F.S. (2003) Detection of staphylococcal enterotoxin B in spiked food samples. *Journal of Food Protection* 66: 1851-1856.

[35] Chang, Y.H., Chang, T.C., Kao, E.F., Chou, C. (1996) Detection of protein a produced by Staphylococcus aureus with a fiber-optic-based biosensor. *Bioscience Biotechnology and Biochemistry* 60: 1571-1574.

[36] Ogert, R.A., Brown, J.E., Singh, B.R., Shriverlake, L.C., Ligler, F.S. (1992) Detection of Clostridium-Botulinum Toxin-a Using a Fiber Optic-Based Biosensor. *Analytical Biochemistry* 205: 306-312.

[37] Narang, U., Anderson, G.P., Ligler, F.S., Burans, J. (1997) Fiber optic-based biosensor for ricin. *Biosensors & Bioelectronics* 12: 937-945.

[38] James, E.A., Schmeltzer, K., Ligler, F.S. (1996) Detection of endotoxin using an evanescent wave fiber-optic biosensor. *Applied Biochemistry and Biotechnology* 60: 189-202.

[39] Masson, J.F., Obando, L., Beaudoin, S., Booksh, K. (2004) Sensitive and real-time fiber-optic-based surface plasmon resonance sensors for myoglobin and cardiac troponin I. *Talanta* 62: 865-870.

[40] Hanbury, C.M., Miller, W.G., Harris, R.B. (1997) Fiber-optic immunosensor for measurement of myoglobin. *Clinical Chemistry* 43: 2128-2136.

[41] Battaglia, T.M., Masson, J.F., Sierks, M.R., Beaudoin, S.P., Rogers, J., Foster, K.N., Holloway, G.A., Booksh, K.S. (2005) Quantification of cytokines involved in wound healing using surface plasmon resonance. *Analytical Chemistry* 77: 7016-7023.

[42] Tschmelak, J., Proll, G., Gauglitz, G. (2004) Sub-nanogram per litre detection of the emerging contaminant progesterone with a fully automated immunosensor based on evanescent field techniques. *Analytica Chimica Acta* 519: 143-146.

[43] Preejith, P.V., Lim, C.S., Kishen, A., John, M.S., Asundi, A. (2003) Total protein measurement using a fiber-optic evanescent wave-based biosensor. *Biotechnology Letters* 25: 105-110.

[44] Tromberg, B.J., Sepaniak, M.J., Vo-Dinh, T., Griffin, G.D. (1997) Fiber-optic chemical sensors for competitive binding fluoroimmunoassay. *Anal. Chem.* 59: 1226-1230.

[45] Hale, Z.M., Payne, F.P., Marks, R.S., Lowe, C.R., Levine, M.M. (1996) The single mode tapered optical fibre loop immunosensor. *Biosensors & Bioelectronics* 11:137-148.

[46] Tang, L., Kwon, H.J., Kang, K.A. (2004) Theoretical and experimental analysis of analyte transport in a fiber-optic, Protein C immuno-biosensor. *Biotechnology and Bioengineering* 88: 869-879.

[47] Garden, S.R., Doellgast, G.J., Killham, K.S., Strachan, N.J.C. (2004) A fluorescent coagulation assay for thrombin using a fibre optic evanescent wave sensor (vol 19, pg 737, 2004). *Biosensors & Bioelectronics* 19: 1775-1775.

[48] Lee, M., Walt, D.R. (2000) A fiber-optic microarray biosensor using aptamers as receptors. *Analytical Biochemistry* 282: 142-146.

[49] Wen-xu, L., Jian, C. (2003) Continuous monitoring of adriamycin in vivo using fiber optic-based fluorescence chemical sensor. *Analytical Chemistry* 75: 1458-1462.

[50] Song, J.M., Kasili, P.M., Griffin, G.D., Vo-Dinh, T. (2004) Detection of cytochrome c in a single cell using an optical nanobiosensor. *Analytical Chemistry* 76: 2591-2594.

[51] Cullum, B.M., Griffin, G.D., Miller, G.H., Vo-Dinh, T. (2000) Intracellular measurements in mammary carcinoma cells using fiber-optic nanosensors. *Analytical Biochemistry* 277: 25-32.

[52] Kleinjung, F., Bier, F.F., Warsinke, A., Scheller, F.W. (1997) Fibre-optic genosensor for specific determination of femtomolar DNA oligomers. *Analytica Chimica Acta* 350:51-58.

[53] Zeng, J., Almadidy, A., Watterson, J., Krull, U.K. (2003) Interfacial hybridization kinetics of oligonucleotides immobilized onto fused silica surfaces. *Sensors and Actuators B-Chemical* 90: 68-75.

[54] Liu, X., Tan, W. (1999) A fiber-optic evanescent wave DNA biosensor based on novel molecular beacons. *Anal. Chem.* 71: 5054-5059.

[55] Almadidy, A., Watterson, J., Piunno, P.A.E., Foulds, I.V., Horgen, P.A., Krull, U. (2003) A fibre-optic biosensor for detection of microbial contamination. *Canadian Journal of Chemistry-Revue Canadienne De Chimie* 81: 339-349.

[56] Pilevar, S., Davis, C.C., Portugal, F. (1998) Tapered optical fiber sensor using near-infrared fluorophores to assay hybridization. *Analytical Chemistry* 70: 2031-2037.

[57] Hermanson, G.T.: Bioconjugate Teshniques. Elsevier, San Diego 1996.

[58] Yan, D., Saunders, J.A., Jennings, K. (2002) Kinetics of formation for n-alkanethiolates self-assembled monolayers onto gold in aqueous micellar solutions of C12E6 and C12E7. *Langmuir* 18: 10202.

[59] Zourob, M., Mohr, S., Brown, B.J.T., Fielden, P.R., McDonnell, M.B., Goddard, N.J. (2005) An integrated metal clad leaky waveguide sensor for detection of bacteria. *Analytical Chemistry* 77: 232-242.

[60] Morgan, M.T., Kim, G., Ess, D., Kothapalli, A., Hahm, B.K., Bhunia, A. (2006) Binding inhibition assay using fiber-optic based biosensor for the detection of foodborne pathogens. *Advanced Nondestructuve Evaluation I, Pts 1 and 2, Proceedings* 321-323: 1145-1150.

[61] Shriverlake, L.C., Ogert, R.A., Ligler, F.S. (1993) A Fiberoptic Evanescent-Wave Immunosensor for Large Molecules. *Sensors and Actuators B-Chemical* 11: 239-243.

[62] Chen, Z., Kaplan, D.L., Gao, H., Kumar, J., Marx, K.A. (1996) Molecular assembly of multilayer enzyme: toward the development of a chemiluminescent-based fiber optic biosensor. *Materials Science and Engineering* C4: 155-159.

[63] Kapoor, R., Kaur, N., Nishanth, E.T., Halvorsen, S.W., Bergey, E.J., Prasad, P.N. (2004) Detection of trophic factor activated signaling molecules in cells by a compact fiber-optic sensor. *Biosensors & Bioelectronics* 20: 345-349.

[64] Zourob, M., Goddard, N.J. (2005) Metal clad leaky waveguides for chemical and biosensing applications. *Biosensors & Bioelectronics* 20: 1718-1727.

[65] Wu, S.Y., Ho, H.P., Law, W.C., Lin, C.L., Kong, S.K. (2004) Highly sensitive differential phase-sensitive surface plasmon resonance biosensor based on the Mach-Zehnder configuration. *Optics Letters* 29: 2378-2380.

[66] Tang, L., Ren, Y.J., Hong, B., Kang, K.A. (2006) Fluorophore-mediated, fiber-optic, multi-analyte, immunosensing system for rapid diagnosis and prognosis of cardiovascular diseases. *Journal of Biomedical Optics* 11: 21011-21011 to 21011-21010.

[67] Hong, B., Kang, K.A. (2006) Biocompatible, nanogold-particle fluorescence enhancer for fluorophore mediated, optical immunosensor. *Biosensors & Bioelectronics* 21:1333-1338.

[68] Minunni, M., Tombelli, S., Mascini, M., Bilia, A., Bergonzi, M.C., Vincieri, F.F. (2005) An optical DNA-based biosensor for the analysis of bioactive constituents with application in drug and herbal drug screening. *Talanta* 65: 578-585.

[69] Thompson, V.S., Maragos, C.M. (1996) Fiber-optic immunosensor for the detection of fumonisin B-1. *Journal of Agricultural and Food Chemistry* 44: 1041-1046.

In: New Developments in Optics Research
Editor: Matthew P. Germanno

ISBN: 978-1-60324-505-7
© 2012 Nova Science Publishers, Inc

Chapter 3

NEW CHALLENGES IN RAMAN AMPLIFICATION FOR FIBER COMMUNICATION SYSTEMS

P.S. André[1,2], A.N. Pinto[1,3], A.L.J. Teixeira[1,3], B. Neto [1,2], S. Stevan Jr.[1,3], Donato Sperti[1,3,4], F. da Rocha[1,3], Micaela Bernardo[2,5], J.L. Pinto[1,2], Meire Fugihara[1,3], Ana Rocha[1,2] and M. Facão[2]

[1]Instituto de Telecomunicações, Aveiro Portugal
[2]Departamento de Física, Universidade de Aveiro, Aveiro, Portugal
[3]Departamento de Electrónica, Telecomunicações e Informática,
Universidade de Aveiro, Aveiro, Portugal
[4]Università Degli Studi di Parma, Parma, Italy
[5]Portugal Telecom Inovação SA, Aveiro, Portugal

ABSTRACT

Raman fiber amplifiers (RFA) are among the most promising technologies in lightwave systems. In recent years, Raman optical fiber amplifiers have been widely investigated for their advantageous features, namely the transmission fiber can be itself used as the gain media reducing the overall noise figure and creating a lossless transmission media. The introduction of RFA based on low cost technology will allow the consolidation of this amplification technique and its use in future optical networks.

This paper reviews the challenges, achievements, and perspectives of Raman amplification in optical communication systems. In Raman amplified systems, the signal amplification is based on stimulated Raman scattering, thus the peak of the gain is shifted by approximately 13.2 THz with respect to the pump signal frequency. The possibility of combining many pumps centered on different wavelengths brings a flat gain in an ultra wide bandwidth.

An initial physical description of the phenomenon is presented as well as the mathematical formalism used to simulate the effect on optical fibers.

The review follows with one section describing the challenging developments in this topic, such as using low cost pump lasers, in-fiber lasing, recurring to fiber Bragg grating cavities or broadband incoherent pump sources and Raman amplification applied to coarse wavelength multiplexed networks. Also, one of the major issues on Raman amplifier design, which is the determination of pump powers in order to realize a specific gain will be discussed. In terms of optimization, several solutions have been published recently, however, some of them request extremely large computation time for every interaction, what precludes it from finding an optimum solution or solve the semi-analytical rate equation under strong simplifying assumptions, which results in substantial errors. An exhaustive study of the optimization techniques will be presented.

This paper allows the reader to travel from the description of the phenomenon to the results (experimental and numerical) that emphasize the potential applications of this technology.

1. INTRODUCTION

The deployment of optical communication systems through long haul networks required the development of transparent optical amplifiers, for replacement of the expensive and limitative optoelectronic regeneration. The increasing distance between amplification sites saves amplification huts reducing by this way the investment and operational cost in the network management.

The first choice for transparent optical amplification pointed out to the Erbium Doped Fiber amplifiers (EDFA), which was a mature technology by the beginning of the last decade of the XX century. However, the growing demand in terms of transmission capacity has been increasing dramatically, fulfilling the entire spectral band of the EDFA, and wideband amplifiers are now required. Raman fiber amplifiers (RFA) have emerged as a key technology for the optical networks.

In lumped amplified systems (using for example EDFAs) the amplification modules are placed every 40~50 km of span. This module amplifies back to the initial power level, the transmission signal attenuated during propagation. The distance between amplifiers is determined by the span loss, by the limit imposed from the maximum admissible power allowed in the fiber without inducing nonlinear effects and by the minimum acceptable power that avois a degradation of the signal-to-noise-ratio.

The use of Raman amplification allows the confinement of the signal inside the limits imposed by the nonlinearities and of the signal-to-noise-ratio degradation resulting from higher span distances. This advantage of the distributed (Raman) over lumped amplification is illustrated in figure 1.

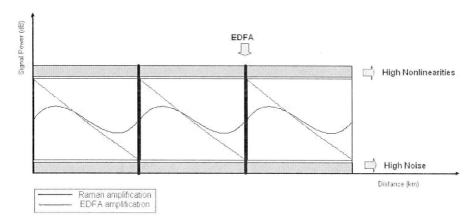

Figure 1. Distributed and lumped amplification signal evolution.

The distributed amplification scheme can be used to cover very long span links or to increase the distance of ultra-long haul systems.

Raman fiber amplifiers are based on the power transfer from pump(s) signal(s) to information carrying signals (usually described as probes) due to stimulated Raman scattering (SRS) which occurs when there is sufficient pump power within the fiber. Since the gain peak of this amplification is obtained for signals downshifted approximately by 13.2 THz (for Silica), relative to the pump frequency, to achieve gain at any wavelength we need to select a pump whose frequency complies with this relation. In this way, it is possible to optimize the number of pumps to obtain a wide and flat gain [1-3]. However, it is necessary to bear in mind, that due to the pump-to-pump interaction, the shorter wavelength pumps demand more power to be effective [4,5].

From a telecommunications point of view, the pump wavelengths must be placed around 1450 nm because the signal wavelengths used on the so called 3rd transmission window are centered around 1550 nm and the maximum gain occurs for a Stokes frequency shift of 13.2 THz.

The RFA had become attractive just after the development and commercialization at a reasonable cost of a key component: the high power pump laser [6]. Typically a high power laser for Raman amplification, provides an optical power of 300 mW, launched over an optical fiber, which for a standard single mode fiber (SMF) is equivalent to a power density of 3.75 GW/m^2. This high power injected into the fiber, especially when multipump lasers are utilized, imposes new concerns in terms of safety.

Therefore, the use of RFAs requires the utilization of automatic power reduction or automatic laser shut down systems to prevent the hazard of high power leakage from the optical cables or service cabinets. Also, as the optical power rises, the nonlinear effects, such as the fiber fuse effect, start to become relevant. This effect has threshold intensity of 10~30 GW/m^2 and it is responsible by a catastrophic destruction of the fiber core. This destruction once started propagates in direction to the optical source, resulting also in the destruction of the pumping laser [7]. For operating wavelengths of 1550 nm, the fuse effect power threshold is ~1.5 W for SMF fibers, while for dispersion shift fibers (DSF) this power is reduced to ~1.2 W [8]. This effect is also responsible by the damage of the optical connectors interface [8].

In terms of implemented systems, several architectures have been proposed, based in all Raman or hybrid Raman/EDFA amplification [10]. The use of bidirectional Raman amplification has also been reported for long reach access networks. Experimental results have shown the feasibility of systems with symmetric up-and-downstream signals with bitrates up to 10 Gb/s, supported by distributed Raman amplification over 80 km of fiber [11]. Field transmission experiments have been reported with 8 × 170 Gb/s over 210 km of single mode standard fiber, achieving spectral efficiency of 0.53 bit/s/Hz [12].

As the traffic increases, wavelength division multiplexing (WDM) arises to enlarge the transmission capacity. This, in turn, requires flexible and broadband architectures which reinforces the interest in Raman amplification. Nowadays, WDM exists in two formats: Dense WDM (DWDM) working at C and L spectral windows, allocating a maximum of 150 channels spaced by 0.8 nm [13], and Coarse WDM (CWDM) working at O, E, S, C, and L spectral windows, allocating a maximum of 18 channels spaced by 20 nm [14]. The DWDM solution is extensively used in long haul systems, sending as much information as possible. CWDM is a good solution whenever less information is transmitted over short distances in a less expensive way than DWDM. As CWDM works with far apart signals, it can make use of uncooled distributed feedback (DFB) lasers [15,16] needing multiplexing components with flexible tolerances. However, as the channels in CWDM systems are far apart, optical amplification is still a matter of concern. Traditional EDFA bandwidth (20~40 nm) cannot support the full band of CWDM channels [17].

Other technical solution to amplification of signals is the semiconductor optical amplifier, which presents a low saturation power (around 13 dBm) when compared with other fiber based amplifiers, but with a signal-to-noise ratio degradation quite considerable. A good solution for the amplification of both DWDM and CWDM relies on Raman amplifiers. A wide and flat spectral gain profile is achievable thanks to the combination of several pumping lasers operating at specific powers and wavelengths. The composite amplification is determined from the mutual interactions among the pump and signal wavelengths. Gain spectra as large as 100 nm were obtained using multiple pumps. Emori et al. have presented an experimental Raman amplifier with a 100 nm bandwidth using a WDM laser diode unit with 12 wavelengths ranging from 1405 to 1510 nm, whose maximum total power was equal to 2.2 W [4, 18]. Therefore, a gain equal to 2 dB is obtained over a 25 km SMF link and a 6.5 dB gain using a 25 km DSF link, both with 0.5 dB of maximum ripple. Kidorf et al. provided a mathematical model to implement a 100 nm Raman amplifier using low power pumps with maximum power of each pump equal to 130 mW [14]. They used 8 pumps from 1416 nm to 1502 nm along 45 km of SMF, obtaining a gain around 4 dB with a maximum ripple equal to 1.1 dB.

The growing maturity of high pump module technologies is providing competitive - solutions based on Raman amplification and currently many alternative techniques are being developed to overcome the ordinary one pump and dual pumping methods [19, 20]. In particular, we report here two major techniques. First, the use of low power pumping lasers provides gain comparable to the ordinary one pump Raman amplification. This technique is especially interesting for combining commercial and low cost lasers [21]. The second particular technique corresponds to an evolution of the cascaded Raman amplification. Actually, a sixth order cascade Raman amplifier was recently proposed [22]. In the cascade Raman amplification, the pump power is downshifted in frequency by using a pair of fiber Bragg gratings (FBG) placed in spectral positions multiples of 13 THz, from the pump frequency. In a particular case, the generation of the fiber pump laser is obtained by using only one passive reflector element and distributed reflectors over the long optical fiber, established by a nonlinear fiber intrinsic effect called Rayleigh backscattering.

The enlargement of the bandwidth of Raman amplifiers is also achieved using incoherent pumping instead of multi-pump schemes [23-27]. Vakhshoori *et al.* proposed a high-power incoherent semiconductor pump prototype that uses a low-power seed optical signal, coupled into a long-cavity semiconductor amplifier. It was achieved 400mW of optical power over a 35nm spectral window [27]. A 50 nm bandwidth amplifier was obtained with an on/off gain equal to 7 dB. It was also demonstrated that the use of six coherent pumps is less efficient, in terms of flatness, than the use of two incoherent pumps [24]. The signal wavelengths were comprised between 1530 nm and 1605 nm and the transmission occurs over 100 km of optical fiber. Another advantage of using incoherent pumping is the reduction of nonlinear effects, such as Brillouin scattering, four wave mixing of pump-pump, pump-signal and pump-noise [28].

RFAs have become a crucial component for the implementation of fiber optic communication systems [9]. An exponential increase on the product distance × capacity of the transmission experiments on optical communication systems was observed in the last decade. The majority of these experiments, especially since the year 2000, have employed RFA as amplification technology [9]. This survey attempts to cover the most recent aspects in the field of Raman amplification for fiber communication systems.

2. THEORETICAL DESCRIPTION OF RAMAN SCATTERING

In 1928 Raman scattering was discovered independently and almost simultaneously by two research groups, one working in India and lead by Sir C. V. Raman [29], and the other by G. S. Landsberg and L. I. Mandelstam working in Russia [30]. In 1930, the Nobel committee distinguished Sir C. V. Raman for his discovery of the molecular scattering of light and since then this effect has been known as the Raman effect.

Raman effect is a scattering effect of light. Light scattering occurs as a consequence of fluctuations in the optical properties of a medium. In optical fibers three types of scattering effects are relevant: Rayleigh, Brillouin and Raman scattering.

Rayleigh scattering is an elastic process, i.e., the incident and the scattered photon have the same energy, therefore the same frequency. Rayleigh scattering in fibers couples light from guided modes to unguided ones leading to optical attenuation. Indeed, in modern fibers operating in the near infrared, Rayleigh scattering is the major source of attenuation, as absorption is practically negligible. In fact, Silica lattice and electronic resonances are in the mid infrared and in the ultra-violet, respectively. Therefore in the near infrared, fibers operate, essentially in an off-resonance regime, apart from impurities, which in nowadays fibers are reduced to an extremely low level [31]. However, besides the off-resonant interaction with bound electrons, optical waves also interact with molecules inside Silica fibers, through scattering.

Raman and Brillouin scattering are both inelastic processes, i.e., the incident and scattering photons have different energies. The energy lost by the incident field is stored into the medium in the form of vibrational energy, named phonons. Indeed, the origin of both Raman and Brillouin scattering effects resides in the interaction of light with these vibrational states (phonons). In the Brillouin scattering low frequency vibrational states are involved, usually referred as acoustic phonons. In the Raman process high frequency vibrational states are presented, named as optical phonons.

Raman scattering can occur in two distinguished forms: Spontaneous Raman Scattering, and Stimulated Raman Scattering (SRS).

In the spontaneous form, Raman scattering occurs when the incident field interacts with vibrational modes, mainly excited by thermal effects, of the molecules constituting the medium. From this interaction, it can result another optical phonon, with frequency Ω, and a down shifted optical photon with frequency $v_S = v_0 - \Omega$, or a up shifted photon of frequency $v_A = v_0 + \Omega$ and in this case an optical phonon is annihilated, v_0 is the frequency of the incident signal. As the frequency Ω is related to the normal vibrational modes of the molecules constituents of the medium, by analyzing the scattered light, information about the medium can be retrieved. This is the main idea behind Raman spectroscopy, a widely used technique for materials characterization. In amorphous materials, like Silica, Ω can assume a value belonging to a broad spectral range, starting from zero and going up to 40 THz. Experimentally both down shifted and up shifted frequencies waves have been observed and have been named as Stokes and anti-Stokes, respectively.

Stimulated Raman Scattering was discovered by E. J. Woodbury and W. K. Ng, almost accidentally in 1962, when working with a Ruby laser [32]. They observed a strong spectral line not coincident with any spectral line of the fluorescence spectrum of Ruby. To understand this process let us assume that an incident photon is scattered by an optical phonon in the medium, and in this process a down shifted photon and an optical phonon are created. We can see that we have two ways of creating phonons, the scattering process and the thermal mechanism. If the intensity of the incident light is small, the rate of phonons created by scattering is low and due to thermal equilibrium the density of phonons in the medium is unchanged, and therefore the medium maintains the same optical properties. If the intensity of light is increased above a certain threshold, the optical properties of the medium can be changed in a way that the scattering process is enhanced [33]. In this situation, the incident light stimulates the scattering process and we are in the presence of Stimulated Raman Scattering. Through this positive feedback the scattering process can be enhanced by several orders of magnitude. Due to the Bosonic nature of the photons, this process can indeed provide gain. The photon emission process by a scattering center, it can be stimulated by the presence of another photon, and this stimulated emission is the origin of the gain. The term emission is used in this context in a quite abusive way because there is no absorption to a real state, but this process can be treated considering that the scattering photon is initially absorbed to a virtual state and after re-emitted.

If we consider that the decay from the virtual states only occurs spontaneously, the Stokes power grows linearly with the pump power. In the other way, if we consider that the decays from the virtual states must be triggered by another photon, the Stokes power grows exponentially with the pump power. Off course, in reality both spontaneous and stimulated emission occurs. If the photon that triggers the stimulated emission is part of a signal we are in the presence of optical gain, which can be beneficial for optical communication systems [34]. If this photon was initially generated by spontaneous emission we are in the presence of amplified spontaneous emission noise which is usually considered as harmful, at least for telecommunications purposes. The spontaneous emission process always leads to an excess of noise in the system.

The optical gain provided by the Raman process can be completely characterized by the Raman-gain coefficient $g_R(\Omega)$, which is related with the imaginary part of the third-order nonlinear susceptibility. The characterization of the amplified spontaneous emission process requires, besides the Raman-gain coefficient $g_R(\Omega)$, another coefficient named noise spontaneous emission factor $n_{sp}(\Omega)$. However, it turns out that another source of noise must be also considered to characterize the noise in Raman amplifiers. This source of noise arises from Rayleigh scattering. Most of the Rayleigh scattered photons are lost through non-guided modes, but some of them are coupled to the counter-propagating mode. Those photons can be amplified and through another Rayleigh scattering process can appear as extra-noise at the amplifier output. This effect is usually named as double Rayleigh scattering and will be described in more detail in section 4.3.

3. MODELING OF RAMAN AMPLIFIERS

The implementation of RFA, using an optical fiber as gain medium, requires that the pump and information signals must be injected into the same fiber. A basic scheme for a RFA architecture is displayed in figure 2. The signal and pump waves are launched into the optical fiber (the gain medium) by a coupler, so, that stimulated Raman scattering can occur. Since the SRS effect occurs uniformly for all the orientations between pumps and signals, Raman amplifiers can work both in forward and/or backward pumping configuration.

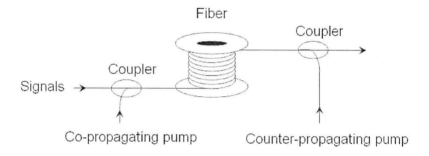

Figure 2. General scheme for a distributed Raman amplifier. For simplicity the optical isolators used to protect the pumps and signals sources, were omitted.

The model for power evolution in Raman amplifiers assuming a multipump multisignal configuration is often based on an unified treatment of channels, pumps and spectral components of the amplified spontaneous emission (ASE). The major interactions can be reasonably drawn by considering the pump-to-pump, signal-to-signal and pump-to-signal power transfer, attenuation, Rayleigh back scattering, spontaneous Raman emission and its temperature dependence. Other effects, such as noise generation due to spontaneous anti-Stokes scattering, polarization and nonlinear index are neglected, but they can reach considerable importance in certain regimes of transmission. It must be noted that signal channels and Raman pumps are treated as fields at single frequencies, so ignoring the interactions due to the spectral shape of signals and pumps.

In a general approach, the power evolution of pumps, signals and ASE (in forward and backward directions), with time along the fiber distance is given by the following set of coupled differential equation [35]. For N_p pumps, N_s probe signals and N_{ASE} spectral components for ASE, the system is formed by $N_p+N_s+2N_{ASE}$ equations.

$$\pm \frac{\partial P_i^\pm(z,t)}{\partial z} \mp \frac{1}{V_i} \frac{\partial P_i^\pm(z,t)}{\partial t} =$$

$$\left[-\alpha_i + \sum_{j=1}^{i-1} g_{ji} \left[P_j^+(z,t) + P_j^-(z,t) \right] - \sum_{j=i+1}^{m} \frac{\upsilon_i}{\upsilon_j} g_{ij} \left[P_j^+(z,t) + P_j^-(z,t) \right] - 2h\upsilon_i \sum_{j=i+1}^{m} g_{ij} \Gamma \left(1 + \eta_{ij} \right) \Delta\upsilon \right] P_i^\pm(z,t) + \gamma_i P_i^\mp(z,t)$$

$$+ h\upsilon_i \sum_{j=1}^{i-1} g_{ji} \Gamma \left[P_j^+(z,t) + P_j^-(z,t) \right] \left(1 + \eta_{ji} \right) \Delta\upsilon$$

$$\tag{1}$$

V_i is the frequency dependent group velocity. The $\pm$ signs stand for the forward or backward propagating waves, being α_i and γ_i the coefficients of attenuation and Rayleigh of the ith wave at frequency υ_i. h and k_B are the Planck and Boltzmann constants, respectively, and T is the fiber absolute temperature. The phonon occupancy factor is given by:

$$\eta_{ij} = \left[\exp\left(\frac{h(\upsilon_i - \upsilon_j)}{k_B T} \right) - 1 \right]^{-1} \tag{2}$$

The frequencies υ_i are numbered by their decreasing value (the lower order corresponds to the higher frequency). Thus, the terms in the summation in expression 1, from $j=1$ to $j=i-1$ cause amplification since the wave i is receiving power from the lower order waves (with higher frequency). For the same reason, the terms in the summation from $j=i+1$ to $j=m$ originate depletion. For mathematical convenience the gain spectrum was divided into slices of width $\Delta\upsilon$, spanning the range over which ASE spectral components are significant.

The terms that contain a product of powers describe the coupling via stimulated Raman Scattering, being its strength determined by the Raman gain coefficient of the fiber, g_{ij} obtained by equation 3.

$$g_{ij} = \frac{g_R(\upsilon_i - \upsilon_j)}{\Gamma A_{eff}} \tag{3}$$

where A_{eff} is the effective area of the fiber and the factor Γ is a dimensionless quantity comprised between 1 and 2 that takes into account the polarization random effects. The achieved gain, as well as the slope of the gain spectrum, depends on the transmission fiber [36, 37]. In figure 3, two Raman gain coefficient spectra are displayed, showing the different strengths of the Raman coupling of a SMF fiber and a dispersion compensating fiber (DCF).

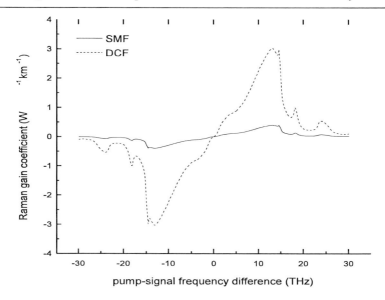

Figure 3. Raman gain coefficient spectra for two germanosilicate fibers: Single mode fiber (SMF) and dispersion compensating fiber (DCF), for a pump wavelength of 1450 nm.

As a matter of fact, the small effective area of the DCF (15 µm²) is determinant for its higher Raman gain coefficient when compared to the SMF (80 µm²) or when compared with DSF fibers (50 µm²). Those spectra also show peaks that are broader than those presented by crystalline materials, since the amorphous nature of Silica allows a continuum of molecular vibrational frequencies.

To obtain a steady-state power distribution, the time derivative in equation 1 is settled equal to zero, and the set of equation takes the form of expression 4.

$$\pm\frac{dP_i^{\pm}}{dz} =$$
$$\left[-\alpha_i + \sum_{j=1}^{i-1} g_{ji}\left[P_j^+ + P_j^-\right] - \sum_{j=i+1}^{m} \frac{\upsilon_i}{\upsilon_j} g_{ij}\left[P_j^+ + P_j^-\right] - 2h\upsilon_i \sum_{j=i+1}^{m} g_{ij}\Gamma\left(1 + \eta_{ij}\right)\Delta\upsilon \right]P_i^{\pm} + \qquad (4)$$
$$+ \gamma_i P_i^{\mp} + h\upsilon_i \sum_{j=1}^{i-1} g_{ji}\Gamma\left[P_j^+ + P_j^-\right]\left(1 + \eta_{ji}\right)\Delta\upsilon$$

In spite of the simplification, the modeling is still computationally intensive, especially for the situation of backward or bidirectional pumping. In those situations, the mathematical problem that describes the power evolution of pumps and signals along the fiber is a boundary value problem (BVP) which is more difficult to solve than the initial value problem (IVP) in the forward pumping scheme. An immediate approach to the numerical solution of such problem is the shooting method [38]. There are other allowable numerical methods, such as relaxation methods, or collocation methods [39]. Generally, shooting methods are faster than relaxation ones. In shooting methods, we choose values for all the dependent variables at one boundary, solve the system of ordinary differential equation (ODE) as an IVP and verify if the obtained values on the other boundary are consistent with the stipulated values (boundary conditions) [40]. Then, the parameters are repeatedly changed using some correction scheme until this goal is attained. The selection of the correction scheme is crucial for stability and efficiency of the resulting algorithm. An other variant of the shooting method, we can guess boundary values at both ends of the domain, integrate the equation to a common midpoint and repeatedly adjust the guessed boundary values so that the solution tends to the same value at the middle point. This adjustment task is usually performed by the Newton-Raphson method.

Recently, some shooting algorithms with different correction schemes for the design of Raman fiber amplifiers have been proposed in order to improve convergence of the solutions even for larger fiber lengths [41]. This scheme is obtained by modifying the numerical method used to perform the IVP integration (fourth order Runge-Kutta, Runge-Kutta-Felhberg, etc). Other approaches to solve the equation 4 propose a shooting method to a fitting point using a correction scheme based on a modified Newton approach. Therefore, by introducing the Broydens rank-one method into the modified Newton method, the algorithm becomes more efficient and stable. This happens because the intensive numerical calculations of the Jacobi matrix are substituted by simpler algebraic calculations [41].

The use of projection methods such as collocation, gives a continuous approximation of the solution as a function of the fiber length. The basic idea is to approximate the BVP solution by a simpler function that represents an approximation.

Nevertheless, the Raman equations (equation 4) are also solvable through semi-analytical methods, using the average power analysis (APA) presented by Min et al. [42]. The amplifier is split into n small segments, in order to avoid the position dependency of the powers of equation 4. The equations are then solved analytically in each segment, considering as input conditions the outputs provided by the solution on the previous segment. Equations 5 to 8 show how the powers are iteratively computed. The output pump/signal power at each section end is given by:

$$P_{out}^{\pm} = P_{in}^{\pm} G(z, \upsilon) \tag{5}$$

being $G(z, \upsilon)$ the section gain,

$$G(z, \upsilon) = \exp\left[(-\alpha(\upsilon) + A(\upsilon) - B(\upsilon))\Delta z\right] \tag{6}$$

The constants, $A(\upsilon)$ and $B(\upsilon)$ are obtained through:

$$A(\upsilon) = \sum_{j=1}^{i-1} g_{ij} \overline{P}_j \tag{7}$$

$$B(\upsilon) = \sum_{j=i+1}^{m} g_{ji} \overline{P}_j$$

The optical power term in each section can be substituted by its length averaged values given by:

$$\overline{P} = P_{in}^{\pm} \frac{G(\upsilon) - 1}{\ln(G(\upsilon))} \tag{8}$$

For a RFA, the net gain is usually defined as the ratio between the signal powers at the end and at the beginning of the fiber link, as defined in equation 9:

$$G_{net} = \frac{P_{signals}(z = L)}{P_{signals}(z = 0)} \tag{9}$$

The so-called on/off gain is another useful quantity that measures the increase in signal powers at the amplifier output when the pumps are turned on, as follows:

$$G_{on/off} = \frac{P_{signals}(z = L) \text{with pumps on}}{P_{signals}(z = L) \text{with pumps off}} \tag{10}$$

The numerical issues due to the backward pumping can be surpassed by assuming that the pump inputs are located at the same fiber end that the signal inputs. Therefore, the pump equations are integrated reversely as if they were backward by multiplying them by (−1). A guessed initial input is necessary to perform the integration, but the algorithm is able to adjust it using an optimization routine that adjust the initial input until the output at the fiber end reaches the real backward pump input. The use of the APA approach has shown a reduction of two orders of magnitude in the computation time, being the obtained results in agreement with the ones resulting from traditional numerical methods.

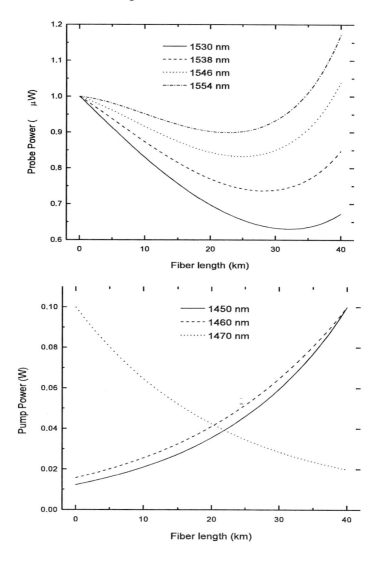

Figure 4. Spatial evolution of two counterpropagated pumps, one copropagate pump and four probe signal along a 40 km SMF fiber span amplifier. Probe signals evolution (top) and pump signals evolution (bottom).

To demonstrate the numerical resolution of the steady-state Raman propagation equations, we assume the scheme in figure 4, where three bidirectional pumps (two backward and one forward) and four probe signals are considered. The counterpropagated pumps have power levels set equal to 0.1W, working at 1450 nm and 1460 nm, respectively. The copropagated pump is working at 1470 nm with an output power also equal to 0.1W. The forward pumping signal are then injected into 40 km of SMF fiber and combined with 4×1000 GHz spaced C band probe signals with an initial optical power equal to 1μW. The spatial evolution of pumps and probe signals are displayed in figure 4.

The implementation of equation 4 also allows the calculation of the total noise for each signal (forward and backward ASE) within the amplifier, whose spatial evolution for this system can be followed in figure 5.

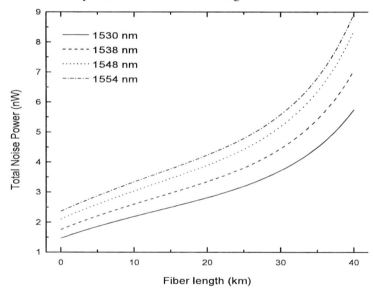

Figure 5. Spatial evolution of the total noise (forward and backward ASE) power along 40 km SMF fiber span.

The noise figure of an optical amplifier amounts the degradation of the signal to noise ratio (SNR) when the signals are amplified. The most important source of noise in optical amplifiers is ASE, which, for Raman amplifiers is due to spontaneous scattering. Assuming that the signals are initially as noiseless as possible, and that their degradation is due to signals spontaneous beat noise produced by ASE, the noise figure, in linear units, is given by equation 11 [36]:

$$NF \approx \left(\frac{2P_{ASE}^+(z=L)}{h\upsilon\Delta\upsilon} + 1 \right) \frac{1}{G_{net}} \tag{11}$$

where $h\upsilon$ is the photon energy and P_{ASE}^+ is the forward ASE measured over the reference bandwidth $\Delta\upsilon$. The first term corresponds to the noise from the signal spontaneous beating and the second one to shot noise.

Another quantity, named effective noise figure, accounts the noise that a discrete amplifier placed at the end of an unpumped fiber link would need to have the same noise performance that a distributed Raman amplifier. In decibel units, the effective noise figure is computed using:

$$NF_{eff}^{dB} = NF^{dB} - (\alpha L)^{dB} \tag{12}$$

Typically, WDM systems allocate a large number of channels spaced over wide bandwidths. Considering the previous system but doubling the pump powers and using 64 probe signals (instead of 4), we obtained the spectra of the gain and noise figure which are plotted in figure 6.

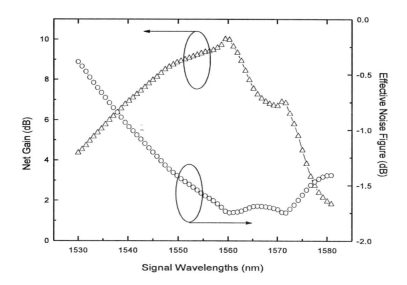

Figure 6. Net gain and effective noise figure spectra for a system with two counterpropagated pumps, one copropagate pump and 64×100 GHz probe signals along 40 km SMF fiber span.

As depicted in this section and despite some remaining numerical issues, the modeling of a multipump Raman amplifier anticipates many valuable applications for WDM systems, namely the broadband gain. It is important to notice that gain spectra as wide as 100 nm are achievable and that the gain value can be kept quite constant by an appropriate tailoring of the amplifier architecture. This procedure involves solely the proper dimensioning of the pump power levels and operating wavelengths, as discussed more extensively in section 4.

Another interesting feature of RFA is the noise performance. The ASE noise in RFA is intrinsically low (as suggested by the negative effective noise figure presented above). The reason relies in the fast relaxation of the optical phonons, the absorption of signal photon to the upper virtual state is extremely small. The inversion of population is almost complete.

4. CHALLENGES IN RAMAN AMPLIFICATION

4.1. Gain Profile Optimization

One of the most impressive features of Raman fiber amplifiers is assuredly the possibility to achieve gain at any wavelength, by selecting the appropriate pump wavelength. Therefore, it is possible to operate in spectral regions outside the Erbium doped fiber amplifiers bands over a wide bandwidth (encompassing the S, C and the L spectral transmission bands). Nevertheless, some studies have been reporting that despite the Raman gain dependence is essentially due to the pump-signal frequency difference; there is also some weaker dependence on the pump absolute frequency [43]. However, since a deeper study of this topic is beyond the scope of this work, we will not consider it in the gain optimization.

A flat spectral gain profile is achievable with the combination of several pumping lasers operating at specific powers and wavelengths. The Raman gain created by pumps at different frequencies is slightly shifted from each other to partly overlap and form a composite gain. When the pump powers and frequencies are properly chosen, this wide gain can also be considerably flat. Another important feature to take into account when designing a flat gain scheme, is the strong Raman interaction between the pumps, since the higher frequency pump is responsible for the amplification of the lower frequency signals, more pumping power is needed, as some will also be transferred to the lower frequency pumps. This interaction between pumps also affects the noise properties of the amplifier. However, some novel pumping schemes have been recently proposed in order to prevent those unwanted effects: copumping, time dependent Raman pumping, higher order pumping and broad-band pumping [44].

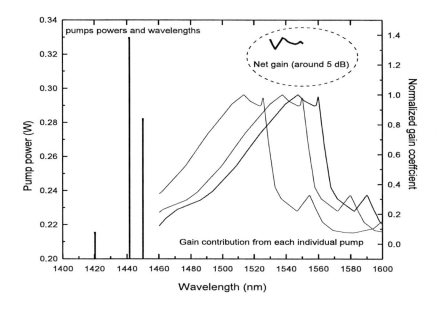

Figure 7. Numerical simulation of broadband Raman amplifier gain. Bars show backward input pump powers and wavelengths. Ticker line show 14×400 GHz probe signals optimized net gain and thin lines the gain contribution of each individual pump. The simulation was carried out through 25 km of SMF fiber.

Typically, laser diodes with output powers in the 100-200 mW range can be used in a multipump scheme. This scheme is normally composed of a set of laser diodes operating in the 14XX nm region, whose spectral width is narrowed and stabilized by a FBG. Optical couplers are used to combine and depolarized them, in order to suppress the polarization dependent gain. The multipumping allows bandwidth upgradeability by the addition of new laser diodes. Theoretically, the larger the number of pumps the better the gain ripples. Nevertheless, there are economic issues that prohibit the use of an arbitrary number of pumps. For this reason, we have to find a balance between the system performance and the cost of amplification.

Optimization of the gain spectrum has been widely performed making use of several global search methods, such as neural networks [45], simulated annealing [46] and genetic algorithm (GA) [47]. During the search process, the pump powers and frequencies are directly substituted into the system of propagation equations to calculate de gain profile. Depending on the speed of the numerical method used to integrate the system of equations, the amount of numerical computations involved can be considerably large and the optimization inevitably time consuming. Those solutions are not suitable for practical applications where the real optimal solution must be provided in a short time.

However, some alternatives can be found by replacing the usual intensive numerical integrations with simpler algebraic calculations using the APA method while integrating the Raman propagation equations.

Another simple but important issue when using a global optimization method relies in a proper dimensioning of the search domain. Using the APA method, all the inputs are located at the same fiber end, even for the counter pump situations. Therefore, the pump power inputs are chosen by presuming a typical propagation profile. By this way, it is advisable to try lower power values for the higher frequency pumps and higher power values for the lower frequency pumps (the opposite happens at other fiber end). Regarding to the optimization of the pump frequencies, it is advisable to divide our spectral range into the number of pumps and then shift those values by 13 THz.

A second approach to speed up the search of the optimal pump configuration uses the genetic algorithm (GA) method only to search the pump frequencies and a quadratic programming method to solve the power integral [48]. The search domain of the GA method is by this way reduced to a half, enabling faster convergence.

Another approach combines GA with the Nelder-Mead search. This so called hybrid GA can be useful in certain situations for the purpose of saving some function evaluations and consequently to perform the optimization in the least time possible [49]. The hybrid GA follows the routine depicted in Figure 8. Firstly, the initial population, as well as the other GA operators are dimensioned: selection, crossover and mutation. The selection, together with the crossover, is responsible for the bulk of GA processing power. The mutation is an operator that plays a secondary role in the GA. Since, the genetic operators can be performed by different methodologies, it is important to choose the ones that are more adequate to the problem we are dealing with, in order to improve the GA search procedure [50]. It must be noted that if the search space is not large, it can be searched exhaustively and the best possible solution will be probably found. The maximum number of allowed generations is also an important feature because, when carefully chosen, it can save a large number of function evaluations. The Nelder-Mead method uses a simplex in a n-dimensional space, characterized by

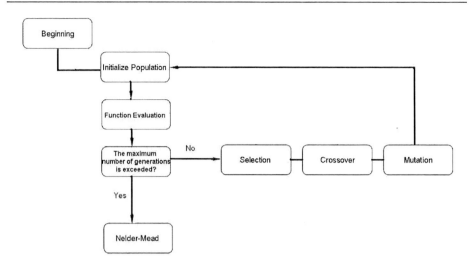

Figure 8. Scheme of the hybrid GA implementation.

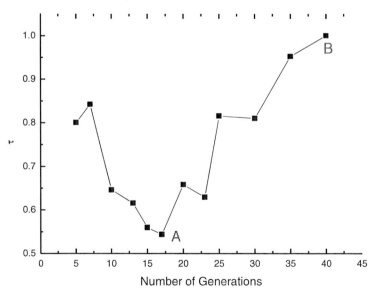

Figure 9. Total simulation time for a hybrid GA against the number of generations for a population size of 50 individuals. The line is a visual guide.

the $n+1$ distinct vectors that are its vertices. At each step of the search, a new point in or near the current simplex is generated. The function value at the new point is compared with the function values at the vertices of the simplex and one of the vertices is replaced by the new point, giving a new simplex. This step is repeated until the diameter of the simplex is less than the specified tolerance. By determining properly the right moment to switch from one method to another, it is possible to reduce the simulation time to a half when compared to simple GA. This result can be observed in figure 9. The normalized total simulation time (τ) against the number of generations is plotted. Here, the reference is the slowest simulation, the one that use 40 generations, identified by the letter B. The best situation (tagged as A) tooks a simulation time equal to about a half of the time of the worse situation and it was attained with 17 generations. An heuristic explanation relies on the intrinsic nature of the GA. We verify that for small number of generations (bellow 15) the GA time is small but the system reaches a worse fittest solution. Thus, the Nelder-Mead method needs more time to reach a desirable solution. When the number of generations increases the GA reaches a best solution but the needed computation time increase accordingly.

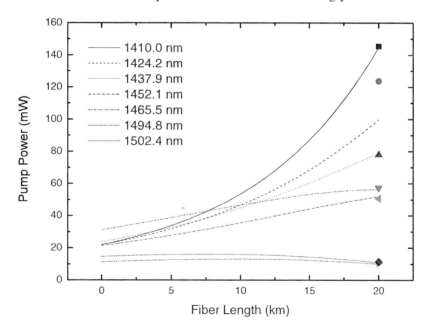

Figure 10. Power evolution of optimized pumps along 20 km of SMF (lines). The geometric shapes stand for the used experimental values.

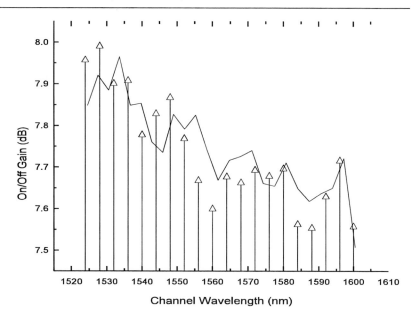

Figure 11. Experimental (arrows) and simulation (line) on/off spectral gain for the 20 probe signals and 7 counter propagated pumps, over 20 km of SMF fiber.

In order to enlighten the conclusions provided by the hybrid GA algorithm, a laboratorial implementation was carried out to test the optimization results. A Raman amplified system with 20 km of SMF fiber, 20 probe signals and 7 backward pumps was implemented. Since the pump wavelengths are already settled, only the optimization of the power levels is needed. The simulation used the stochastic uniform method for selection, the scattered crossover method and the uniform mutation. A population of 50 individuals and a number of generations equal to 35 were considered. The spatial evolution of the pumps signals optimized values are displayed in figure 10 jointly with the pump signal experimental values. In figure 11, the optimized and experimental on/off gain spectra are presented.

This is a good agreement between the optimization modeling and the experiment. The maximum ripple attained by the optimization is 0.41 dB being the experimental maximum ripple equal to 0.23 dB. The mean square deviation between simulation and experimental results is equal to 0.0036. Indeed, a flat gain over a wide bandwidth (~80 nm) was attained, using seven pumps with a total input power equal to 453 mW.

4.2. Raman Amplification Using Multiple Low Power Lasers

One of the main issues in Raman amplification is related to the stability of the high power lasers, the costs and the need for efficient cooling. To go around these problems, the usual solution is the use of several pump signals, what results in added advantages, like high, flat and wide-gain bandwidth [51-53].

The technology evolution allowed that high power pumps are nowadays commercially available, although some problems still limited [54]. The pressure on optical components prices, lead to the creation of CWDM standards [55]. This is reflected specially on price dropping of uncooled lasers with relatively high powers (>10mW). The price to pay is wavelength wondering, however, neither for CWDM nor for Raman, wavelength stability is not a stringent requirement, allowing simple control. With this technology the possibility of achieving Raman gain by combining multiple of these low power lasers was successfully implemented [21].

Teixeira *et el* proposed the use of an array of low cost lasers to achieve wideband Raman amplification, providing both experimental and simulation results [21]. In this work a counterpropagating topology was implemented, using 40 C band lasers with 20 mW output power spaced by 0.8 nm (1533 nm − 1557 nm). These lasers were combined using a multiplexer, bringing up a total power of more than 200mW (23 dBm). This power is enough to generate SRS. Several fibers were tested: True Wave and dispersion compensating. To characterize the gain profile, an array of 40 L-band 0.8nm spaced probe lasers (1565 nm -1605 nm) with a total optical power of 1 mW was used. Figure 12 (a) shows the implemented setup. Figure 12 (b) presents the simulation results for the implemented system to four different pumping configurations. The first curve corresponds to the traditional approach, where one high power pump (23.6 dBm) at a single wavelength (located at 1530 nm) is used. In the second case, three lasers spaced by 0.8 nm starting at 1530 nm having total power of 23.6 dBm were multiplexed. The results for the two above pumping configurations are approximately equal, having only a wavelength shift of 0.8 nm as expected due to the average pumps wavelength difference. Similar simulation was experienced considering 40 lasers, each with 7.6 dBm (after the multiplexer), resulting in a total power of 23.6 dBm. In this case, the gain curve appears even smothered and the peak shifted by ~16 nm; the peak gain is similar, however a small enhancement on the 3 dB gain bandwidth was obtained and the gain profile smothered. In order to explore the advantages of the methodology (gain flatness), the power distribution for the pumps was optimized to reach an equalized gain, while maintaining the same total pump comb. The peak gain was decreased at the expense of an increased flattened profile.

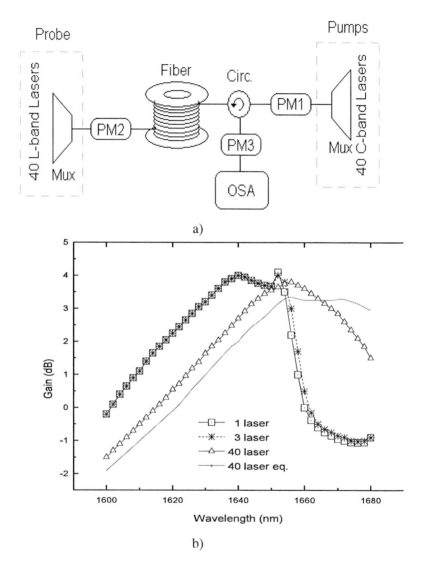

a)

b)

Figure 12. a) Implement setup for the simulation and experimental systems, b) simulated Raman gain profiles for several sets of pumping configurations, with 23.6 dBm of total power. PM demotes an optical power meter, OSA denotes an optical spectrum analyzer and MUX is an optical multiplexer.

Due to limitations on available probe and pump signals, the experimental implemented system only can scope part of the spectral bands used in simulation. The gain is only measurable when the pump powers go above 10 dBm. A maximum of 3 dB net gain was achieved in the L band for full pump power, 23.6 dBm, as displayed in figure 13 b). Also, in the same figure, a minimum of 2dB gain over more than 30nm, with 1dB ripple, was achieved without any power distribution optimization. The results have demonstrated the effectiveness of the technique to achieve Raman amplification.

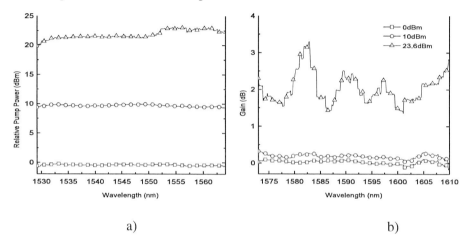

a)   b)

Figure 13. a) Pump to pump Raman effect; b) experimental Raman gain achieved for several values of the pump power, with probes at 0 dBm.

In figure 13 (a) it is illustrated the pump to pump effect which is commonly occurring in dense WDM systems. This effect starts to be noticeable above 10 dBm of total power and is evident for 7.6 dBm per channel. This phenomenon can be harmful due to uneven distribution of power during transmission, however, if correctly considered can be used to obtain beneficial extra gain in the system, if a pre equalization is also implemented .

4.3. Raman Amplification Using Rayleigh Backscattering

Raman amplification pumping can also be achieved by recurring to the traditional methods of shifted gain [19, 20]. In these methods, several FBG reflector pairs are used to generate resonant cavities in the maximum of the Raman gain spectrum. Thus, with a Ytterbium laser operating in the vicinity of

1090 nm, where it exhibits its maximum efficiency, it is possible to generate pumps in the E band, as demonstrated by Papernyi *et al*, where a set of 6 FBG reflector pairs were used to generate pumping in the E-band [22]. The latter amplifies the C band, where the probe signals transmission usually occurs.

The main penalties of traditional Raman amplification are associated with intrinsic nonlinear phenomena such as nonlinear refraction and Rayleigh backscattering, since it is required to use high powers and long fiber spans. This last effect occurs when a fraction of scattered light is backreflected towards the launch end of the optical waveguide. This reflection is called single Rayleigh backscattering (SRB). Part of this scattered light is also backreflected in the forward direction and it is called double Rayleigh backscattering (DRB), as shown in figure 14 [56]. SRB and DRB can be controlled by actuating properly on the fiber drawing process or by a correct power design [57]. The Rayleigh backscattering has been studied, modeled and characterized by many authors [56-59]. It is known that the process results from multiple reflections of light inside the fiber and therefore spontaneous and unstable lasing can occur [60]. However, this phenomenon has been observed as an impairment to signal transmission [61, 62].

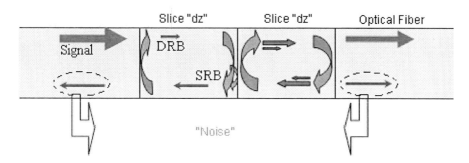

Figure 14. Simple Rayleigh backscattering (SRB) and double Rayleigh backscattering (DRB) representations over an infinitesimal length of fiber.

Recently, a method that, up to some extent, allows the control of this phenomenon was reported [56,63]. With the possibility of controlling the SRB and DBR effect, novel applications can be drafted. One suggestion is the use of this effect to generate distributed resonant cavities, which will degenerate in lasing if enough gain is achieved. These are achieved with the help of only one end FBG set [63]. This is advantageous when compared to the previously described methods to obtain cascaded Raman amplification, since it needs only one FBG set, minimizing the need for identical FBG to be used and tuned at different sites which can be not colocated.

In order to demonstrate the application of this technique to control SRB and DRB, the experimental system reported in figure 15 was implemented. A Raman pump in the E-band, at 1428 nm, was coupled to the transmission fiber, with controllable power up to 1.5 W. A circulator was used to protect the laser from back reflections and, simultaneously, to allow the measurements of the back reflected power spectrum. Two different scenarios were observed: the FBGs are absent between the fiber and the coupler; and the setup was complete as described in figure 15. These two scenarios target to show the controlling effect achieved by the FBGs.

A set of three FBG with wavelengths centered at: 1520 nm, 1531.6 nm, 1535.6 nm all having 95% reflectivity, were placed after the pump and act as reflective elements.

In a first setup, a 14 km DCF fiber with dispersion parameter equal to -1393 ps/nm and Raman coefficient of 3.05×10^{-3} m^{-1}W^{-1} was used as transmission medium.

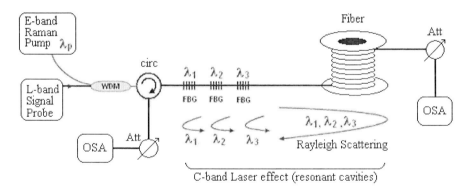

Figure 15. Experimental setup for the double shifted Raman experiments; WDM denotes a band coupler and Att denotes an optical attenuator.

Considering the first scenario, where no FBGs were present, the common Raman effect in the C band was observed, figure 16 a) for a pump power of 350 mW. When the power of the pump was increased to 600 mW, Rayleigh backscattering spontaneous lasing effect is observed, as displayed in figure 16 b). This effect presents random behavior, both in wavelength and power, being the spectrum time dependent.

In a second scenario the FBGs were present, control of the random process generated by the SRB and DRB was achieved and the lasing was stabilized in the FBGs wavelengths. In this situation a virtual cavity was established, formed by the FBG and the Rayleigh backscattered light. To generate more than one laser in the C-band a set of cascaded wavelength mismatched FBGs were used. These gratings are responsible for a multipeak frequency dependent reflection back into the fiber of the amplified spontaneous emission and DRB light from the fiber. This, in conjunction with the FBG, create resonant cavities, which generate stable wavelength constant lasing actions, from now on called as FBG-DRB lasing.

Due to the different reflectivities of the FBGs and the Raman gain profile, different lasing powers for each configuration occur in the C-band. Whenever the power of the generated lasers in the C-band is high, cascaded Raman effects will occur that generate gain in the far L and U-band. The FBG-DRB lasing and consequent stabilization process with the simultaneous L-U band spontaneous emission is reported in figure 16 c), where a pump power of 1.2 W was used [63].

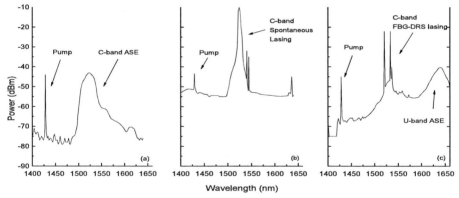

Figure 16. Transmission spectra for 14 km DCF fiber: a) Spontaneous ASE for a pump of 300mW; b) spontaneous lasing for a pump of 600mW; c) C-Band FBG-DRB lasing and far L and U-band Raman generated ASE for a 1.2 W pump.

The results show a 38 nm flattened ASE bandwidth in the U-band, generated by the FBG-DRB. By introducing a copropagating probe at 1625 nm, a gain of 10 dB was measured for an E-band pump power of 1 W.

In a second setup, different optical fibers were tested in order to compare the pump power laser threshold. A 14 km long DCF fiber, a 50 km long DSF fiber and a 50 km long non zero dispersion shift fiber (NZDSF) were used [60]. Figure

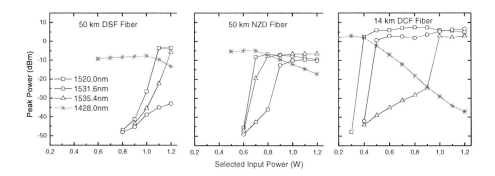

Figure 17. Depletion of the E-band pump and peak power of the C band lasers as a function of the pump power for several fiber types; from left to right: DSF, NZD and DCF.

17 shows the different lasing thresholds and curve shapes resulting from the intrinsic differences between the optical fibers. From figure 17, it can be observed that this process is more efficient in the DCF fibers, where the threshold power is 350 mW, while for the NZDSF fiber is 650mW and 1W for the DSF fiber.

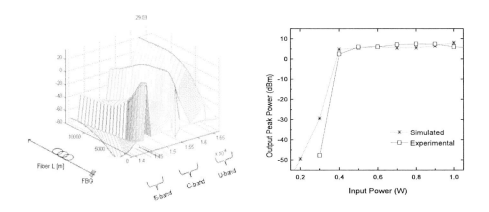

Figure 18. (a) Optical power density spectra for 14 km DCF fiber from E-band to U-band; (b) Output power evolution of the lasing effect of the FBG-DRB at 1520 nm.

Usually, the simulation of Raman amplification as convergence and stability problems, especially for high pump powers, has reported in previous sections. The simulation of the lasing effect with high pump power has similar difficulties. To avoid such problems the solving method for the differential equation system is simplified to an analytical method based on the transfer matrix (APA) as proposed in section 4.1. Inside these fiber slices, the parameters are considered to have small variations and the solution of the equation system is obtained by stabilization after multiple passes along the length of the fiber [64]. The initial solution uses an analytical approach that was based in the undepleted case. The approach to the pump depletion is included in the attenuation of the pump.

In each fiber slice, the Rayleigh backscattering is calculated at the boundary and this backscattering power is added to the signals in the same direction and wavelength, that also suffer amplification and depletion.

Figure 18 (a) presents the simulation results of such algorithm for 29.03 dBm of pump power. The evolution of the power densities from E-band to U-band spectra is shown for a long fiber span. The E-band pump signal suffers depletion in the long propagation fiber. This pump works as a seed of the C-Band FBG-DRB lasing, which generate the L-U-Band Raman gain.

Since the process of lasing is not stable, the simulation process presents a slow stabilization, but, the boundary powers over the FBG are quickly stabilized. Figure 18 (b) presents a comparison of the threshold laser power obtained by experimental and semi-analytical methods. The output peak power of the FBG-DRB signal at 1520 nm is related with the input pump power.

As observed for the DCF fiber, stable multiple laser actions were achieved for moderate pump powers (350 mW) for both simulation and experiment.

4.4. Amplification with Incoherent Pumps

A technique to increase the bandwidth and decrease the spectral ripple of RFA is available with incoherent pump lasers. A Raman amplifier with incoherent pumps can be modeled as a multipump Raman amplifier. In such case, the spectrum of the incoherent pump is well approximated by a large number of pumps of infinitesimal spectral width and whose power sum equals the integral power of the incoherent pump. Therefore, the theoretical model used for incoherent pump schemes is based on the model, previously presented, for coherent multipump configurations.

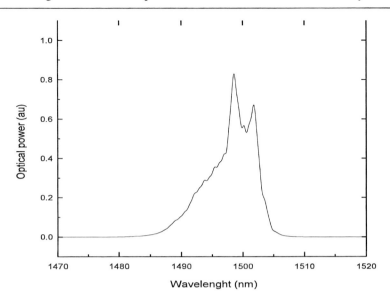

Figure 19. Pump spectrum for the incoherent pump.

An incoherent pump spectrum, as displayed in figure 19, with 10 nm FWHM, can be approximated by 100 pumps of infinitesimal spectral width, having an aggregate power equal to the integral power of the incoherent pump. The incoherent pump here considered was obtained from a high power FBG (Fiber Bragg Grating) laser, from which the stabilization grating was removed [65].

To evaluate the advantages of this technique, the Raman on/off gain and the noise figure were measured for coherent and incoherent pumping over 40 km of SMF fiber. The probe signal combo consists of 13 channels, with 1 mW power, spaced by 100 GHz over the 1546-1556 nm spectral region. Both co-propagating and counter-propagating architectures were considered. The coherent pumping source was a high power FBG laser with a wavelength of 1490 nm. In both cases the pump power was 290 mW.

The results of Raman on/off gain and effective noise figure are shown in figure 20. The relatively low on/off gain is due to the fact that the pump wavelengths have not been optimized for this signal band.

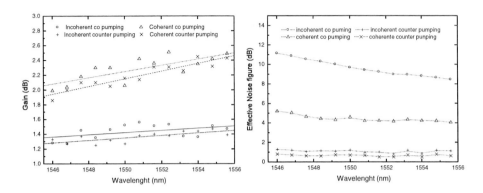

Figure 20. Raman gain and Effective Noise Figure. Lines are simulated results and points represents to experimental data.

The incoherent pumping gain slopes are 0.015±0.008 dB/nm and 0.017±0.004 dB/nm for co and counter propagation configurations, respectively. For coherent pumping, the gain slopes are 0.042±0.01 (co-propagation) and 0.052±0.005 dB/nm (counter-propagation). Such results show that the incoherent pumping configuration presents a flatter gain.

The noise figure is approximately the same for coherent and incoherent pumping in the counter-propagating configuration. However, in the co-propagating case, the noise figure is considerably lower for coherent pumping.

In agreement with previous works [23-26], these results indicate that the incoherent pumping technique can be used to decrease the spectral ripple of the Raman gain.

4.5. Raman in CWDM Systems

Another important challenge is the deployment of RFA for access networks, namely for CWDM networks. Since CWDM systems require large bandwidths to guarantee the transmission of a reasonable number of channels, spaced by 20 nm, wide band Raman amplifiers are well suited for this purpose.

The Raman amplifier bandwidth can be enlarged by using multiple pumps. Optimization of the number of pumps and their wavelengths enables the large needed gain spectra and that could be placed in any range of wavelengths used in optical communications.

The design of an amplifier that fits more than two CWDM channels can be achieved, with the following procedure. The number of channels to be transmitted is determined in order to define the required bandwidth. The optical fiber characteristics impose a minimum to the required gain, and finally the number of pumps as well as their characteristics are decided. The scheme of figure 21 illustrates the important issues to be considered to design a multi-pumped Raman amplifier for a CWDM system.

Since the number of CWDM channels and the length of the link as well as its losses are defined, the minimum required gain to compensate the transmission losses and the minimum bandwidth to transmit all the required channels may be determined using the rectangle shown in figure 21. The gain has to be high enough to compensate the losses caused by the optical fiber and the bandwidth should be large enough to support all the transmitted channels. The purpose is to obtain a spectrum that encloses this rectangle.

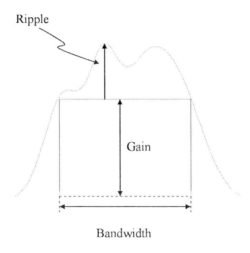

Figure 21. Design concerns for a multi-pumped Raman amplifier for CWDM systems.

Another point is to guarantee that the maximum deviation between the values of the designed and the needed gain as smallest as possible. The curved line in figure 21 represents the obtained spectrum after optimization of pumps characteristics. The ripple represents the maximum deviation cited above.

Another concern in designing the spectrum is to make it flat, with all the channels at the same level, in order to avoid reception constrains.

The multipumped Raman amplifier can be designed using a set of coupled nonlinear equations as equation 4. Solving the coupled equations for one signal and one pump, may be simplified when pump depletion is ignored. This approximation is valid because the pump power is higher than the signal power,

$P_p \gg P_s$ [66]. However, whenever multiple pumps are used this simplification cannot be used due to the interaction between pumps which enhances the effect of depletion due to the higher powers involved.

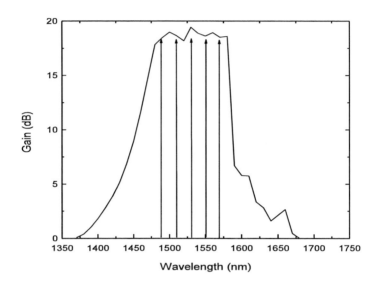

Figure 22. Example of a multi-pumped Raman amplifier applied to CWDM systems.

Figure 22 shows an example of an optimized Raman amplifier spectrum applied to CWDM systems. It was designed to transmit five probe channels at 1490 nm, 1510 nm, 1530 nm, 1550 nm and 1570 nm. The transmission link is based on 80 km SMF fiber, with 0.23 dB/km losses, which implies a 18.4 dB gain with a minimum spectral bandwidth of 80 nm.

The graph in figure 22 is the result of a forward pumping configuration. The number of pumps used in this example was six. The continuous line represents the gain spectrum obtained with the six pumps the arrows represent the transmitted probe channels.

The bandwidth is 100 nm, 20 nm larger than the minimum required bandwidth, in order to guarantee that all the signals are amplified. The gain is around 18.4 dB with a maximum deviation between the designed and needed gain equal to 1 dB, and a maximum gain deviation for each channel being 0.9 dB.

The six pumps used are centered at 1380 nm, 1393 nm, 1405 nm 1428 nm, 1444 nm, and 1468 nm with powers of 450 mW, 200 mW, 330 mW, 160 mW, 45 mW, and 55 mW, respectively. The pump of lower wavelength needs the highest power due to the interactions between pumps: The lower wavelength pump loses energy to the higher wavelengths, causing its depletion.

This optimization scheme was verified experimentally, with 3 probe channels CDWM system, pumped with 3 pump signals at 1470 nm, 1490 nm and 1510 nm. For the optimization the hybrid GA algorithm, previously presented, was used. This pump allocation problem is less exigent, in terms of ripple, than for a DWDM system, since the probe signal are far apart.

The implemented scenario consists of a 40 km SMF fiber, with a counterpropagating pump scheme and a 7 dB gain target. The optimized pump powers were 128.1 mW, 65.0 mW and 146.9 mW, respectively. The maximum gain excursion was 0.002 dB and 0.12 dB for the simulation and experimental systems, respectively.

Experimentally, we can observe that the Raman amplification improves the eye opening penalty of a signal transmitted along a fiber link allowing a good reception at the end of the transmission path. To illustrate this behavior, the eye diagram of a signal after a 40 km link is shown in figure 23.

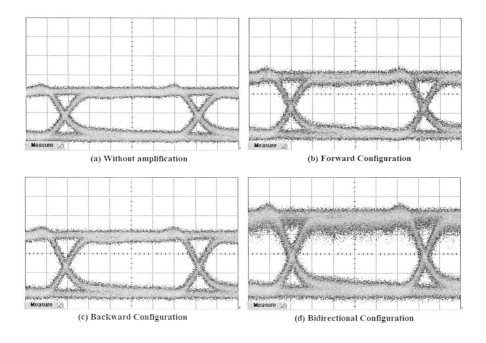

(a) Without amplification   (b) Forward Configuration

(c) Backward Configuration   (d) Bidirectional Configuration

Figure 23. Comparison between eye diagrams with and without Raman amplification.

Figure 23 illustrates a real case where there is a signal centered at 1567 nm and two pumps centered at 1508.8 nm, one in forward configuration and another in backward configuration. The powers of the pumps are chosen to be 100 mW each.

The eye openings are given in Volt and the gain was obtained using the on/off definition (equation 10). Using the relation between power and voltage, $P=V^2/R$, the on/off gain becomes $G_{Voltage} = 10\log_{10}(V_{with\ pump}/V_{without\ pump})$.

It is notorious that an increase of the eye opening obtained when both pumps are turned on. The scale is the same to all the graphs in figure 23 to allow comparisons of the eye opening amplitude. The eye opening to the bidirectional configuration is higher due to the higher pump power, while the forward and backward systems have 100 mW, the bidirectional system uses 200 mW. The respective gains of the eye openings are 1.94 dB, 2.35 dB, and 2.98 dB to the forward, backward and bidirectional systems, respectively. The results show a higher gain for the counter propagating situation.

5. CONCLUSION

Raman fiber amplifiers are a technological key component that fulfill the challenging strict requirements of the beginning of this century, enabling applications not feasible with conventional EDFAs.

In this contribution, we have discussed the origin of Raman scattering and the critical properties for system design, such as pumping allocation, cascade pump and broadband amplification for multiple CDWM networks. It was also presented solutions that provide that gain, such as the use of low power pumps or incoherent pumps.

These issues are, in the authors point of view, the relevant questions and challenges associated with Raman amplification on communication systems.

Acknowledgments

This work was supported by the POSC program, financed by the European Union FEDER fund and by the Portuguese scientific program. The authors also greatly acknowledge the ARPA (POSI/EEA-CPS/55781/ 2004) and TECLAR (POCI/A072/2005) projects and to FCT and ALBAN scholarship program.

REFERENCES

[1] Hu, J.; Marks, B. S.; Menyuk, C. R. *J Lightwave Technol.* 2004, 22, 1519-1522.

[2] Giltrelli, M.; Santagiustina, M. *IEEE Photon Technol Lett.* 2004, 16, 2454-2456.

[3] Cui, S.; Liu, J.; Ma, X. *IEEE Photon Technol Lett.* 2004, 16, 2451-2453.

[4] Kidorf, H.; Rottwitt, K.; Nissov, M.; Ma, M.; Rabarijaona, R. *IEEE Photon Technol Lett.* 1999, 11, 530-532.

[5] Bromage, J. *J Lightwave Technol.* 2004, 22, 79-93

[6] Namiki, S.; Seo, K.; Tsukiji, N.; Shikii, S. *Proceedings IEEE* 2006, 94, 1024-1035.

[7] Shuto, Y.; Yanagi, S.; Asakawa, S.; Kobayashi, M.; Nagase, R. *IEEE Journal Quantum Electronics* 2004, 40, 1113-1121.

[8] Seo, K.; Nishimura, N.; Shiino, M.; Yuguchi, R.; Sasaki, H. *Furukawa Review* 2003, 24, 17-22.

[9] Bromage, J. *J Lightwave Technol.* 2004, 22, 79-93.

[10] Chen D. Z., Wellbrock G., Peterson D. L., Park S. Y., Thoen E., Burton C., Zyskind J., Penticost S. J., Mamyshev P., *ECOC* 2006, Cannes.

[11] Monroy I. T., Kjaer R., Palsdottir B., Koonen A: M. J., Jeppesen P., *ECOC* 2006, Cannes.

[12] Scheiders, M.; Vorbeck, S.; Leppla, R.; Lach, E.; Schimidt, M.; Papernyi, S.; Sanapi, K, *Optical fiber conference* 2005, Anaheim, Post-deadline paper.

[13] ITU -T G.694.1, 2002.

[14] ITU -T G.694.2, 2002.

[15] Eichenbaum, B. R.; Das S. K. *National Fiber Optic Engineers Conference* 2001, Baltimore, USA, 1444-1448.

[16] Wang, D.; He, C.; Li, Y. *FibreSystems Europe/Lightwave Europe.* 2005, 13-15

[17] Desurvire E. *Erbium-Doped Fiber Amplifiers - Principles and Applications*; John Wiley & Sons, New York, NY: 1994.

[18] Emori, Y.; Tanaka, K.; Namiki, S. *Electron Lett.* 1999, 35, 1355-1356.

[19] D. I. Chang, H. K. Lee and K. H. Kim, *Electron. Lett.* 1999, 35, 1951-1952.

[20] S.B.Papernyi,V.I.Karpov,andW.R.L.Clements, *Optical Fiber Communication (OFC)* 2002, Anaheim, USA, PostdeadlinePaperFB4-1.

[21] Teixeira, A.; Andre, P.; Stevan, S.; Silveira, T.; Tzanakaki, A.; Tomkos, I. *International Conference on Internet and Web Applications and Services/Advanced International Conference AICT-ICIW* 2006, on 19-25

Feb. 2006 Page(s):85 - 85 ,Digital Object Identifier 10.1109/AICT-ICIW.2006.158

[22] Papernyi, S.B., Ivanov, V.B. ,Koyano, Y., Yamamoto, H., "Sixth order cascaded Raman Amplification .- Ivanov", *Optical Fiber Communication Conference*, 2005. Technical Digest. OFC/NFOEC

[23] Han, B.; Zhang, X. P.; Zhang, G. D.; Lu, Z. G.; Yang, G. X. *Opt Express.* 2005, 13, 6023-6032.

[24] Zhang, T.; Zhang, X.; Zhang, G. *IEEE Photon Technol Lett.* 2005, 17, 1175-1177.

[25] Wen, S. F. *Opt Express.* 2006, 14, 3752-3762.

[26] Wen, S. F. *Opt Express.* 2006, 15, 45-55.

[27] Vakhshoori, D.; Azimi, M.; Chen, P.; Han, B.; Jiang, M.; Knopp, K. J.; Lu, C. C.; Shen, Y.; Vander Rhodes, G.; Vote, S.; Wang, P. D.; Zhu, X. *OFC.* 2003, 3, PD47-P1-3.

[28] X. Zhou, M. Birk and S. Woodward, *IEEE Photon. Technol. Lett* 2002, *vol* 14, 1686-1688.

[29] C. V. Raman, K. S. Krishnan, "A New Type of Secondary Radiation", *Nature*, Vol. 121, pp. 501-502, March, 1928.

[30] G. S. Landsberg, L. I. Mandelstam, "Eine neue Erscheinung bei der Lichtzerstreuung in Krystallen", *Naturwissenschaften*, Vol. 16, pp. 557-558, July, 1928.

[31] P. S. André, A. N. Pinto, *Chromatic Dispersion Fluctuations in Optical Fibers Due to Temperature and Its Effects in High-Speed Optical Communication Systems, Optics Communications*, Vol. 246, Issues 4-6, 15 February 2005, pp. 303-311, 2005;

[32] E. J. Woodbury, W. K. Ng, "Ruby Laser Operation in the Near IR", *Procedings of IRE* (correspondence), Vol. 50, No. 11, pp. 2367, November, 1962.

[33] R. W. Boyd, Nonlinear Optics, second edition, San Diego, Academic Press, 2003;

[34] C. Headley, G. P. Agrawal, *Raman Amplification in Fiber Optical Communication Systems*, San Diego, Academic Press, 2005.

[35] Karásek M., Kanka J., Honzátko P., Peterka P., *Int. J. Numerical Modelling: Electronic Network, Devices and Fields* , 2004, vol. 17, nº2, 165-176.

[36] Bromage J., *J. Lightwave Technol*, 2004, vol. 22, nº1, 79-93.

[37] André, P., Teixeira, A., Kalinowsky, H., Pinto, J. L. *Optica Aplicatta* 2003, 33
559 – 573.

[38] Liu X., Lee B., *Opt. Express*, 2003, vol. 11, nº.12, 1452-1461.

[39] Neto B.; Stevan S; Teixeira A. T; André P. S.; *ICT* 2006, Funchal.

[40] Han Q., Ning J., Chen Z., Shang L., Fan G., *J. Opt A Pure Appl. Opt.,* 2005, 7, 386-390.

[41] Han Q., Ning J., Zhang H., Chen Z., *J. Lightwave Technol,* 2006, 24, 1946-1952.

[42] Min B., Lee W. J., Park N., *IEEE Photon. Technol Lett.,* 2002, 12, 1486-1488.

[43] Newbury N. R., , *J. Lightwave Technol,* 2003, 21, 3364-3373.

[44] Mollenauer L. F., Grant A. R., Mamyshev P. V., *Opt. Lett.,* 2001, 592-594.

[45] Xiao P. C., Zeng Q. J., Huang J., Liu J. M, *IEEE Photon. Technol Lett.* 2003, 15,
206-208.

[46] Yan M., Chen J., Jiang W, Li J., Chen J., Li X., *IEEE Photon. Technol Lett.* 2001, 13, 948-950.

[47] Zhou X., Lu C., Shum P., Cheng T. H., *IEEE Photon. Technol Lett.* 2001, 13, 945-947.

[48] Cui S, Liu L, Ma X, *IEEE Photon. Technol Lett.* 2004, 16, 2451-2453.

[49] Neto, B., Junior, S., Teixeira, A., André, P. *European Conf. on Networks and Optical Communications - NOC* 2006, Berlin, Germany.

[50] Goldberg D. E., *Genetic Algorithms in Search, Optimisation and Machine Learning,* Massachusetts: Addison-Wesley co, 1989, pp 28-56

[51] V. E. Perlin and H. G. Winful *J. Lightwave Technol.* 2002, 20, 250–254.

[52] X. Liu, J. Chen, C. Lu, and X. Zhou, *Opt. Express* 2004, 12, 6053-6066.

[53] X. M. Liu, et al., *J. Lightwave Technol.* 2003, 21, 3446-3455.

[54] M. Islam, *IEEE Journal of Selected Topics in Quantum Electronics* 2002, 8.

[55] T. Miyamoto, R. Lindsay, *Lightwave Magazine* 2003, 1.

[56] Kobyakov, A., Gray S. and Vasilyev M. *Electronics Letters* 2003, 39, 732 – 733.

[57] Tsujkawa K., Tajima K., Ohashi, M. *J. Lightwave Technol.* 2000, 18, 1528-1532.

[58] Essambre R.; Winzer P.; Bromage J.; Kim C. H., *Photonics Tech. Letters* 2002, 14,
914 – 916.

[59] Park, J., Kim N. Y., Choi W.; Lee, H., Park N. *Photonics Tech. Letters* 2004, 16, 1459 – 1461.

[60] Teixeira A., Stevan Jr., S. Silveira T.; Nogueira R.; Tosi Beleffi G. M., Forin D., Curti F. *European Conf. on Networks and Optical Communications - NOC* 2005,London, UK.

[61] P. B. Hansen, L. Eskildsen, A. J. Stentz, T. A. Strasser, J. Judkins, J. J. DeMarco, R. Pedrazzani, and D. J. DiGiovanni, *IEEE Photon. Technol. Lett.* 1998, 10, 159-161.

[62] Faralli, S. Di Pasquale, F., *IEEE Photonics Technology Letters* 2003, 15, 804- 806.

[63] S. Stevan Jr., A. Teixeira, T. Silveira, P. André, G. M. Tosi Beleffi, A. Reale and A. Pohl, Double shifted Raman amplification by means of spontaneous Rayleigh Backsattering lasing control, *ITC* 2006.

[64] S. Stevan Jr. , A. Teixeira, P. André, G. M. Tosi Beleffi, A. Pohl, simulation of Raman amplification and Rayleigh Scattering laser using the transference matrix method, *MTPT* 2006, Leiria, Portugal.

[65] André, P., A. N. Pinto, Teixeira, A.T., Neto, B., Junior, S., Spertti, D., Rocha, F., Bernardo, M., Fujiwara, M., Rocha, A., Facão, M. *ICTON* 2007, Rome, Italy.

[66] Agrawal, G. P. *Nonlinear Fiber Optics*, 3rd ed. San Diego: Academic Press, 2001.

In: New Developments in Optics Research   ISBN: 978-1-60324-505-7
Editor: Matthew P. Germanno   © 2012 Nova Science Publishers, Inc

Chapter 4

FIBER BRAGG GRATINGS IN HIGH BIREFRINGENCE OPTICAL FIBERS

Rogério N. Nogueira, Ilda Abe and Hypolito J. Kalinowski

Instituto de Telecomunicacoes, polo de Aveiro,
Aveiro, Portugal

ABSTRACT

Fiber Bragg gratings (FBG) are a key element in optical communication devices and in fiber sensors. This is mainly due to its intrinsic characteristics, which include low insertion loss, passive operation and immunity to electromagnetic interferences. Basically a FBG is a periodic modulation of the core refractive index formed by exposure of a photosensitive fiber to a spatial pattern of ultraviolet light in the region of 244–248 nm. The lengths of FBGs are normally within the region of 1–20 mm. Usually a FBG operates as a narrow reflection filter, where the central wavelength is directly proportional to the periodicity of the spatial modulation and to the effective refractive index of the fiber. The production technology of these devices is now in a mature state, which enables the design of gratings with custom-made transfer functions, crucial for all-optical processing. Recently, some work has been done in the application of FBG written in highly birefringent fibers (HiBi). Due to the birefringence, the effective refractive index of the fiber will be different for the two transversal modes of propagation. Therefore, the reflection spectrum of a FBG will be different for each polarization. This unique property can be used for advanced optical processing or advanced fiber sensing.

The chapter will describe in detail this unique device. The chapter will also analyze the device and demonstrate different applications that take advantage of its properties, like multiparameter sensors, devices for optical communications or in the optimization of certain architectures in optics communications systems.

1. INTRODUCTION

The development of the fiber optical technology was an important step in the revolution of global communications and in information technology. One of these developments happened in the 70's with the first optical fibers with low attenuation [1], a feature that enabled long- distance communication with high bandwidth. The intrinsic optical bandwidth of the optical fibers has also allowed the propagation of different simultaneous channels, allowing the transmission of data at Tbit/s rates [2]. In these systems, in addition to transmission and amplification, it is often necessary to do all-optical processing to the signal. This is due to the inherent advantages of the optical processing, relative to the optic-electric-optic processing, like the higher flexibility to operate at different bit rates and modulation formats and also at the higher bandwidth. The evolution of the fiber optical technology has also enabled the development of devices for all optical processing. In this way, the insertion loss is reduced and the processing quality improved. One of the factors contributing to all-fiber optical processing devices was the discovery of the photosensitivity in optical fibers. It was documented for the first time in 1978 by Hill et al. [3] and led to the development of fiber Bragg gratings (FBG).

A FBG is, generally speaking, a periodic perturbation, along the longitudinal axis, of the refractive index in the fiber core. The production of the refractive index perturbation is done optically in a photosensitive fiber. With the current techniques, it is possible to produce fiber Bragg gratings with different optical properties, which can be designed according to the desired optical processing. In addition to the high flexibility in the production of gratings with custom amplitude and phase responses, the compatibility with common transmission fiber also reduces the insertion loss and decreases the production costs.

The application in optical sensors is also a large potential market for FBG. Their intrinsic low immunity to electromagnetic interference, high dynamic range, passive operation, resistance to corrosion and the possibility of multiplexing hundreds of sensors have made FBG a quite interesting sensor for different applications including medicine, civil, aeronautics or biomechanics. Their properties enable the measurement of temperature and also deformation with extremely high resolution. Nevertheless, it can also be used to measure other parameters using indirect measurements [4-7]. The high potential of these devices has also induced the creation of several companies dedicated to the production and installation of fiber sensors.

There are already good references for the study of FBGs [8,9]. The purpose of this chapter is not to study in detail these devices, but to describe a special case when a fiber Bragg grating is written in high birefringence fibers (HiBi FBG). These special gratings have unique polarization properties that give them exclusive capabilities for optical communications. This is due to the possibility of applying a different optical processing for different polarization components of the signal being transmitted.

HiBi FBGs are also quite interesting for multiparameter sensors, due to their response to temperature variations and deformation. Sensors capable of measuring simultaneously several physical parameters have increased in importance in today's technological world. In particular, there are various applications of such sensors in civil, mechanical, biomedical or aeronautical engineering, where measurements of different parameters are required [10]. Engineering structures are an example of an application area for the multiparameters sensors, where strain sensing can lead to better understanding about their lifetime and failure. Such knowledge can be critical for some applications like smart skins for airplanes and aeronautical vehicles.

2. FIBER BRAGG GRATINGS

A FBG is an optical device produced within the core of a standard optical fiber (figure 1). Basically, it is a periodic modulation of the core refractive index

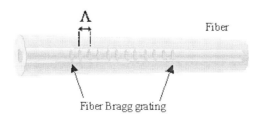

Figure 1. Scheme of a Fiber Bragg grating written in an optical fiber.

formed by exposure of a photosensitive fiber to a spatial pattern of ultraviolet light. The length of a FBG is dependent on its application, but it generally varies between a few millimeters to a few centimeters.

The periodic modulation of the refraction index generates a resonant condition at the Bragg's wavelength (λ_B) which is given by the Bragg's condition:

$$\lambda_B = 2n_{eff}\Lambda \tag{1}$$

where n_{eff} is the effective refraction index of the fiber and Λ is the modulation period. Therefore, when a FBG is illuminated by a broadband source, a spectral band centered at λ_B will be reflected back. The reflection function can be determined using the coupled mode theory [11-14], since it is difficult to determine analytically. The exception is the uniform FBG, where it is possible to calculate the reflectivity in an analytical way. Considering a uniform periodic modulation of the refractive index, with amplitude Δn, the reflection coefficient of the grating can be given by

$$\rho(\lambda) = \frac{-\kappa \sinh(\varphi L)}{\delta \sinh(\varphi L) + i\varphi \cosh(\varphi L)} \tag{2}$$

where L is the length of the FBG, the propagation constant mismatch, δ, is given by

$$\delta = \frac{2\pi n_{eff}}{\lambda} - \frac{\pi}{\Lambda}, \tag{3}$$

$\varphi = \sqrt{\kappa^2 - \delta^2}$, and κ is the coupling constant given by

$$\kappa = \frac{\pi \Delta n}{\lambda} \eta \tag{4}$$

where η is the overlap integral and can be approximated as $\eta \approx 1$ for single mode fibers with step index.

The reflectivity is given by

$$R = |\rho|^2 = \frac{\sinh^2(\varphi L)}{\cosh^2(\varphi L) - \dfrac{\delta^2}{\kappa^2}} \tag{5}$$

and the phase by

$$\phi_R = \arctan\left|\frac{\mathrm{Im}(\rho)}{\mathrm{Re}(\rho)}\right| \tag{6}$$

Figure 2 shows the calculated reflectivity and the phase of a uniform FBG with $L = 5$ mm and $\Delta n = 2 \times 10^{-4}$ as given by the above equations.

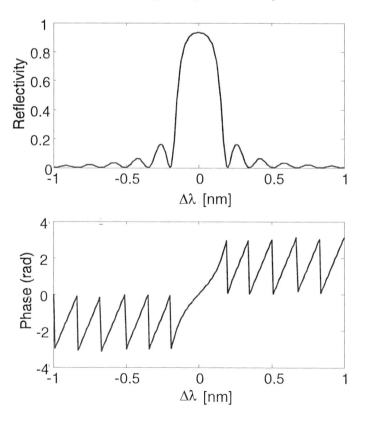

Figure 2. Reflectivity and phase of a uniform FBG. Parameters: L=mm and $\Delta n = 2 \times 10^{-4}$.

If the period changes linearly with the length of the grating, the FBG is said to have a linear chirp. Figure 3 shows the simulation of the reflectivity and group delay of a linear chirped FBG. The simulation method is based on the coupled mode theory.

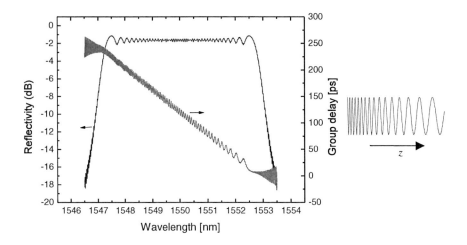

Figure 3. Reflectivity and group delay of a linear chirped FBG.

3. HIGH BIREFRINGENCE FIBERS

In an ideal monomode fiber, with a perfect cylindrical core, and with uniform diameter, the fundamental propagation mode is a degenerated combination of two orthogonal propagation modes. However, in real fibers, that degeneration does not exist. In fact, small variations of diameter in the fiber's core generate a birefringence in the optical fiber. The birefringence can also be a result of an anisotropic stress in the fiber. The local birefringence, B, in each position of the fiber, is defined as

$$B = \left| \overline{n}_x - \overline{n}_y \right| = C_f \left(\sigma_x - \sigma_y \right) \tag{7}$$

where $\overline{n}_x$ and $\overline{n}_y$ are the mean refractive index of the orthogonal polarization modes, σ_x and σ_y are the main stress on the polarization axes and C_f is the photoelastic constant of the fiber. In monomode silica fibers C_f is around

3.08×10^{-6} mm^2/N for wavelengths near 1500 nm, while B is typically $B \approx 10^{-7}$. Due to this small birefringence value, the two polarization components of the light propagating in the fiber have a propagation velocity very similar. Therefore, small environmental perturbations will lead to an energy coupling between one polarization to another. As a result, a linearly polarized light will rapidly evolve to a random polarization. This situation can be avoided with high birefringence fibers. In these fibers, the core has an anisotropic stress, which is generated due to the geometric properties of the fiber. Due to the photoelastic effect, the stress induces a birefringence in the core. Typical values are $B \approx 10^{-4}$ [15]. Due to the high birefringence, the propagation constant is different for the two orthogonal propagation modes, which means that the coupling between both transversal propagation modes is far lower as compared to standard fibers. Therefore, the higher the birefringence, the easier will be for a linearly polarized light, propagating in one of the orthogonal modes, to maintain its state of polarization. Due to this feature, HiBi fibers are also known as polarization maintaining fibers. Figure 4 shows the main structure of the most common HiBi fibers.

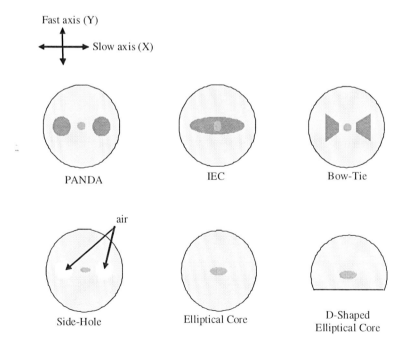

Figure 4. Schematic of the transversal section of some of the most known HiBi fibers.

The PANDA (Polarization-maintaining AND Attenuation-reducing), IEC (Internal Elliptical Cladding) and Bow tie fibers have anisotropic glass structures around the core, with a Poisson coefficient different from the rest of the fiber. These structures create the anisotropic stress in the core, which produces the birefringence. The Side-Hole, the Elliptical Core and the D-Shaped Elliptical Core fibers have an elliptical core to generate the birefringence, aided by two air structures, in the case of the Side-Hole or by the shape of the cladding, in the case of the D-Shaped elliptical core. The main axes of the HiBi fibers are designated as fast axis (Y) for the lower refraction index and slow axis (X) for the higher refraction index.

Coherence Length

If a linearly polarized light propagates in a monomode fiber, with a polarization angle of 45°, relatively to the main axes of the fiber, both orthogonal polarization modes will be excited with equal power. If the fiber has a constant birefringence, the mismatch, $\Phi_{HB}(z)$, between the orthogonal polarization components will change as a function of the propagation distance on the fiber, z, and it's given by

$$\Phi_{HB}(z) = (\beta_x - \beta_y)z \qquad (8)$$

where β_x and β_y are the propagation constants in the X and Y axes respectively. The mismatch will change periodically with the fiber, leading to a change in the state of polarization from linear to elliptical and back again to linear (figure 5).

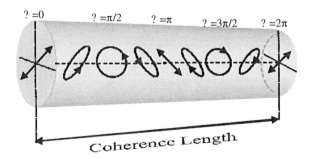

Figure 5. Evolution of the state of polarization in a birefringence fiber.

The spatial periodicity of the evolution of the state of polarization is designated as coherence length (L_B). It is determined by the birefringence of the fiber and can be expressed as

$$L_B = \lambda / B \tag{9}$$

where λ is the operating wavelength. Typical coherence lengths for HiBi fibers are in the millimeter scale [16].

4. FIBER BRAGG GRATINGS WRITTEN IN HIBI FIBERS

HiBi fibers can have two linear polarization modes with refractive indexes n_x and n_y for the slow and fast modes respectively. When a FBG is written in one of these fibers, the periodic modulation will be the same for the two orthogonal polarization modes; however since the effective refraction index is different for the two polarizations, the Bragg wavelength will also be different for each mode. Consequently, expression (1) can be rewritten for the two orthogonal modes:

$$\lambda_i = 2n_i \Lambda \quad , i = X, Y \tag{10}$$

where λ_i are the Bragg wavelengths for each polarization mode.

The wavelength difference between the two reflection peaks, $\Delta\lambda_{HB}$, can be calculated by

$$\begin{aligned}\Delta\lambda_{HB} &= \lambda_x - \lambda_y \\ &= 2n_x\Lambda - 2n_y\Lambda\end{aligned} \tag{11}$$

The reflectivity of a HiBi FBG will be given by the linear sum of the reflectivity of the two polarization components, i.e. $R(\lambda) = R_x(\lambda) + R_y(\lambda)$. R_x and R_y are the reflectivity for each polarization given by

$$R_i = |\rho|^2 = \frac{\sinh^2(\varphi_i L)}{\cosh^2(\varphi_i L) - \dfrac{\delta_i^2}{\kappa^2}} \quad , i = x, y \tag{12}$$

where

$$\delta_i = \frac{2\pi n_i}{\lambda} - \frac{\pi}{\Lambda} \ , \ \mathrm{i=x,y} \tag{13}$$

and

$$\varphi_i = \sqrt{\kappa^2 - \delta_i^2} \ , \ \mathrm{i=x,y} \tag{14}$$

Figure 6 shows a simulation, using the previous model, for the reflectivity of a HiBi FBG with birefringence of $B = 3.2 \times 10^{-4}$.

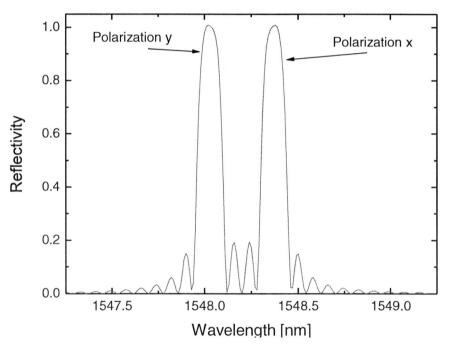

Figure 6. Reflectivity of a simulated HiBi FBG. Simulation parameters: $B=3.2 \times 10^{-4}$, $\Lambda=535$ nm, $L=10$ mm.

If the HiBi FBG is illuminated with light having the two orthogonal components, the reflection spectrum will have those two peaks at orthogonal polarizations. This feature can be very important in some applications, namely in optical communications, as it will be confirmed further in this chapter.

The production of HiBi FBGs uses the same techniques as the ones used in regular FBGs. The only difference will be in the utilization of photosensitive HiBi fiber. Generally it is used a hydrogenated HiBi fiber.

Table 1 shows the dimensions of the anisotropic glass structures around the core of some HiBi fibers obtained through the photographs of the transverse section. The table also displays the main characteristics of HiBi fibers obtained from the manufacturers data sheet.

The reflection spectra for gratings written in the above fibers are shown on figure 7, where the plots of the best-fitted bands are also presented [19]. All the gratings were produced with the phase mask technique. The estimated length of the grating is 10 mm.

From these spectra it can be seen the effect of the intrinsic birefringence of the HiBi fibers. The IEC fiber has the higher birefringence, corresponding to larger spectral splitting between both polarizations bands, while the bow tie fiber presents the lowest birefringence.

Table 2 shows the best-fit parameters obtained in the simulation process. From the fit it is also possible to obtain the values of birefringence of the HiBi fibers.

Table 1. Characteristics of different HiBi fibers. The structures of the HiBi fibers were obtained by microphotography

Fiber type	Commercial provider	Wavelength (nm)	Core diameter (μm)	Cladding diameter (μm)	Intrinsic stress-applying region
IEC (FS-PM-6621)	3M	1300	8	125	Ellipse: Major axis: 75 µm; Minor axis: 30 µm
Bow tie (F-SPPC-15)	Newport	1550	8	125	From core center to extremity of bow tie lobe: 18.4 µm
Bow tie (HB-1500G)	Fibercore	1550	8	80	From core center to extremity of bow tie lobe: 16.5 µm
PANDA (SM-13-P-7)	Fujikura	1300	8	125	From core center to opposite extremity of side cylinder: 41 µm; Diameter of side

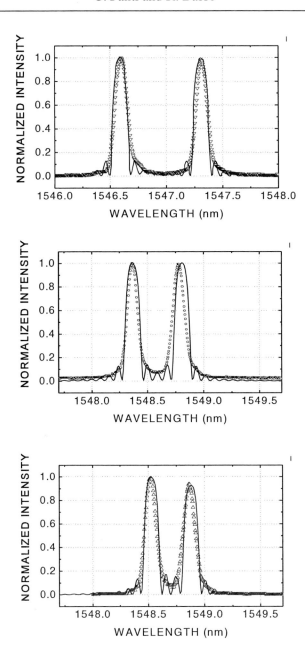

Figure 7. Reflection spectra of Bragg gratings written in different HiBi fibers: IEC (∇), Panda (O) and bow tie (Δ). The continuous line represents the simulated best fit.

Table 2. Parameters of FBGs written in HiBi fibers obtained for the best fit for the experimental data

HiBi Fiber	Bands	λ (nm)	n_{eff}	kL	Λ (nm)	B
IEC	λ_Y	1546.57	1.44539	1.7212	535	6.7 x 10^{-4} @ 1550 nm
(FS-PM-6621)	λ_X	1547.29	1.44606	1.7196		
PANDA	λ_Y	1548.39	1.44709	1.7172	535	4.1 x 10^{-4} @ 1550 nm
(15P8)	λ_X	1548.82	1.44750	1.7162		
Bow tie	λ_Y	1548.61	1.44730	1.7167	535	3.2 x 10^{-4} @ 1550 nm
(SPPC-15)	λ_X	1548.95	1.44762	1.7159		

The Bragg wavelength peaks of the optical spectrum for both polarizations can change with temperature and strain. Therefore, considering a HiBi FBG under a temperature variation of ΔT and under a strain aligned with the main axes of the fiber $\Delta\varepsilon_X$, $\Delta\varepsilon_Y$ and $\Delta\varepsilon_Z$, the resultant wavelength shift, $\Delta\lambda_x$ and $\Delta\lambda_y$ of both wavelength peaks, λ_x and λ_y, can be expressed as

$$\frac{\Delta\lambda_X}{\lambda_X} = \Delta\varepsilon_Z - \frac{n_X^2}{2}[p_{11}\Delta\varepsilon_X + p_{12}(\Delta\varepsilon_Z + \Delta\varepsilon_Y)] + \left[\alpha + \frac{(\partial n/\partial T)}{n_X}\right]\Delta T \quad (15)$$

$$\frac{\Delta\lambda_Y}{\lambda_Y} = \Delta\varepsilon_Z - \frac{n_Y^2}{2}[p_{11}\Delta\varepsilon_Y + p_{12}(\Delta\varepsilon_Z + \Delta\varepsilon_X)] + \left[\alpha + \frac{(\partial n/\partial T)}{n_Y}\right]\Delta T \quad (16)$$

where p_{11} and p_{12} are the components of the photoelastic tensor and α is the thermal expansion coefficient of the fiber, $\alpha = 0.55 \times 10^{-6}$ K^{-1} [17]. For a fiber based on germanium and silica, $p_{11}=0.113$, $p_{12}=0.252$ and the thermo-optic coefficient is $\frac{(\partial n/\partial T)}{n} = 8.6 \times 10^{-6}$ [18].

Figure 8 shows schematically the effect on the reflection spectrum of a HiBi FBG when it is under temperature variations, under transversal strain or longitudinal strain. The effect of temperature variations or longitudinal strain in the reflection spectrum is equivalent to a translation in the wavelength. On the other hand, when under a transversal strain, the peak separation will change. This difference can be used in multiparameter sensors as it will be discussed further in this chapter.

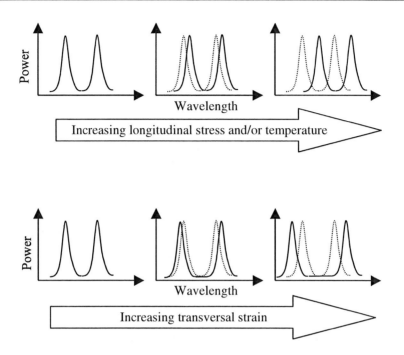

Figure 8. Evolution of the reflection spectrum of a HiBi FBG when under a longitudinal stress, temperature variation or transversal strain.

4.1. Characterization of Bragg Gratins Written in High-Birefringence Fiber Optics

4.1.1. Transverse Strain

The sensitivity of HiBi FBG to transversal strain can be characterized using a mechanical set-up, like the one shown in figure 9. The transversal load is applied using a micro scratch mechanical system. The system uses an arm to apply a load with a precision of 0.1 N. A grating written in HiBi fiber was placed between two plates having a length of 13 mm. The apparatus arm applies the load to the upper plate. The transverse loads were made for several orientations of the birefringence axis with respect to the direction of the applied load through two fiber rotators. Figure 9 also shows the optical system used to analyze the FBG reflection spectrum. Optical spectra were recorded using an amplified spontaneous emission (ASE) of an erbium doped fiber amplifier as light source, an optical circulator and conventional optical spectrum analyzer (OSA).

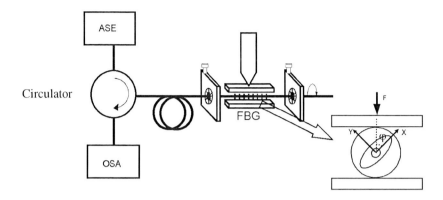

Figure 9. Set-up for the characterization of HiBi FBGs under a transversal load. The detail shows the transverse section of a HiBi fiber oriented along the angle φ.when subjected to applied force F. ASE: Broadband optical source (amplified spontaneous emission); OSA: optical spectrum amplifier.

Figure 10 (a), (b) and (c) shows an example of the reflection spectra of a FBG written in a IEC HiBi fiber as a function of an applied load of 0°, 45° and 90°, respectively.

The results show that, if a load is applied to one of the main axes, fast or slow, it leads to a change in the wavelength of the spectral band associated with the orthogonal axis, while the band associated with the correspondent axis will show a smaller variation. For the applied load angle of 45° both polarization bands present similar evolution. The figure also shows, for different applied load angles (φ), the evolution of the peak wavelength of each reflection band with the transverse strain applied to the sample. The strain calibration points in the spectra deformed areas were obtained by identifying and measuring local maximum, minimum and inflexion points. The band split that occurs in some of the spectra is due to a phase shift induced by the applied load. The resulting complex structure is known to be responsible for spectral changes of FBG subject to mechanical stress [9].

Figure 11 shows the wavelength sensitivity curves obtained for both polarization bands. The graph also displays the periodic evolution of the bands as a function of the applied load angle.

Identifying and measuring the reflection peaks as a function of the applied load can be used to obtain the calibration line for each polarization band. The respective slopes can be evaluated and, from them, the dependence of the Bragg wavelength position with the strain can be obtained. Table 3 shows some measurements obtained with FBGs written in IEC and PANDA fibers.

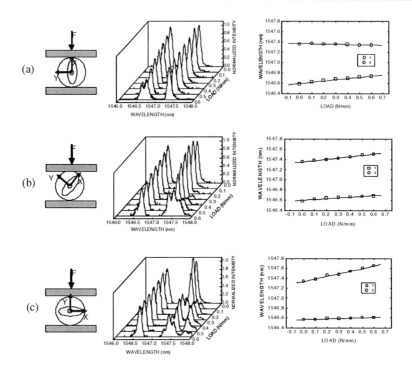

Figure 10. Left: Changes in the spectral response of a FBG written in an IEC HiBi fiber when subjected to an applied load oriented along the angle φ: (a) 0° (X-axis); (b) 45° and (c) 90° (Y-axis). Right: Peak position of each band as a function of the applied load. The lines represent the linear best fit for the experimental data.

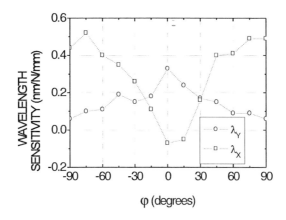

Figure 11. Curves of peak sensitivities of the FBG in IEC HiBi fiber as a function of the applied load angle.

Table 3. Slopes and strain sensitivities of FBGs written in IEC and PANDA HiBi fibers as a function of the direction of applied load (module values). Both fibers have a diameter of 125 μm

HiBi Fiber	Angle of applied load	X- polarization band		Y- polarization band	
		Slope (nm/N/mm)	Strain sensitivity (pm/με)	Slope (nm/N/mm)	Strain sensitivity (pm/με)
IEC	$\varphi = 90°$	0.51	7.02	0.07	1.02
	$\varphi = 0°$	0.02	0.29	0.11	1.55
PANDA	$\varphi = 90°$	0.46	3.78	0.02	0.24
	$\varphi = 0°$	0.01	0.11	0.13	2.80

4.1.2. Longitudinal Strain

The Bragg wavelength dependence with the longitudinal strain can be measured by gluing one extremity of the fiber in a holder, while the other is glued to a translation stage, which applies a known deformation using a calibrated micrometer.

Figure 12 shows the reflection spectra of a FBG and peak position of each band, written in an IEC fiber as a function of longitudinal strain.

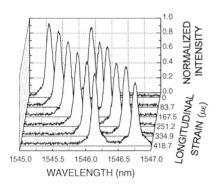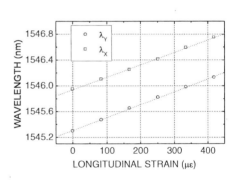

Figure 12. Left: Changes in the spectral response of a FBG written in IEC HiBi fiber when subjected to a longitudinal strain. Right: Peak position of each band as a function of the longitudinal strain. The lines represent the linear best fit for the experimental data.

Both bands show the same behavior, which is an increase of peak wavelengths as the strain increases. The slopes and the Bragg wavelength sensitivity to longitudinal strain are given in table 4. The obtained ratios between strain and applied load were 758 με/N (X-axis) and 755 με/N (Y-axis).

Table 4. Slopes and longitudinal strain sensitivity of a FBG written in an IEC HiBi fiber.

Bands	Slope (nm/N)	Longitudinal strain sensitivity (pm/$\mu\varepsilon$)
X – polarization	1.44	1.9
Y - polarization	1.51	2.0

4.1.3. Temperature

The temperature dependence of the reflection bands of HiBi FBGs can be characterized using a cooling/heating system. Figure 13 shows the evolution of the reflection bands and peak position of each band of a Bragg grating written in an IEC fiber as a function of temperature.

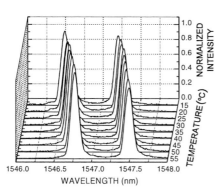 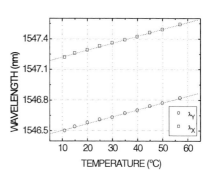

Figure 13. Left: Changes in the spectral response of a FBG written in an IEC HiBi fiber when subjected to different temperatures. Right: Peak position of each polarization band as a function of the temperature. The lines represent the linear best fit for the experimental data.

Table 5. Slopes of temperature for FBGs written in HiBi fibers.

HiBi fiber	Slope (pm/°C)	
	X – polarization band	Y – polarization band
IEC 125 μm	6.76	6.71
PANDA 125 μm	3.28	3.40
Bow Tie 125 μm	10.93	11.12
Bow Tie 80 μm	8.02	8.46

Table 5 shows the temperature sensitivity values for a FBG written in IEC, PANDA and bow tie HiBi fibers.

The results show that there are quite large variations between the sensitiveness to temperature for different HiBi fibers. The values changed between ~ 3 pm/°C for the PANDA fiber to ~ 11 pm/°C for the IEC fiber. There are also differences in the coefficients between polarization bands of the same fiber. For example, in the PANDA fiber this difference is 0.24 pm/°C. These results can be used for simultaneous measurements of temperature and longitudinal strain with only one FBG in a HiBi fiber. This approach, along with others, will be described in the next section.

5. APPLICATION IN MULTIPARAMETER SENSORS

FBG sensors are generally based on a unique grating written in a standard fiber optic. The wavelength shift in the reflection spectrum may be used to measure a single component of strain or temperature variation, but not both simultaneously. An adequate measurement of both temperature and strain requires a suitable sensor with a differential sensitivity between parameters. HiBi FBGs can be used as sensors to simultaneously measure one component of transverse strain, temperature and/or longitudinal strain. As it was shown previously in this chapter there are differences in the calibration coefficients of both polarization bands, which can be used to simultaneous measure the temperature and longitudinal strain with only one HiBi FBG. Since the variations of temperature or longitudinal strain causes both bands to shift, and the variation of strain causes asymmetric spectral response in the polarization bands depending of the direction of the applied load, allows the FBG in the HiBi fiber to measure simultaneously transverse strain and temperature or transverse strain and longitudinal strain.

Several types of optical sensors using FBG written in HiBi fibers, which simultaneously measure longitudinal strain and temperature have been proposed and demonstrated [20-24]. Some of the methods include the recording by a CCD camera of the LP_{01} and LP_{11} spatial modes [22], using a HiBi FBG partially exposed to chemical etching [20] or by using a quasi-rectangular HiBi fiber to increase the birefringence [21]. In those works, only the longitudinal strain component was measured in simultaneous with temperature. But, there are many applications where it is desirable to determine the transverse strain components in addition to longitudinal strain. Several techniques based in HiBi FBG have already been reported for transverse strain sensing [19, 25-29]. However, when a transverse strain is applied to a HiBi FBG, depending of the fiber orientation relatively to the applied load, the separation of the two Bragg wavelengths can be quite low, so it becomes impossible to resolve the two peaks. To overcome this problem, it can be used an interrogation system capable of detecting independently and simultaneously the two orthogonally polarized signals reflected from the HiBi FBG [26].

There are many applications where it is necessary an ultra small sensor to measure simultaneously components of transverse strain, longitudinal strain and temperature. The use of two superimposed Bragg gratings in HiBi fiber have been described in the literature like potential sensors for monitoring four parameters, two components of transverse strain, longitudinal strain and temperature. [30-34].

5.1. Simultaneous Measurement of Transverse Strain and Temperature

The change in the Bragg wavelength of a HiBi FBG, for each polarization, due to a temperature change ΔT and a transversal strain $\Delta \varepsilon$, is given by

$$\Delta \lambda_X = \frac{\partial \lambda_X}{\partial T} \Delta T + \frac{\partial \lambda_X}{\partial \varepsilon} \Delta \varepsilon \tag{17}$$

$$\Delta \lambda_Y = \frac{\partial \lambda_Y}{\partial T} \Delta T + \frac{\partial \lambda_Y}{\partial \varepsilon} \Delta \varepsilon \tag{18}$$

were $\partial \lambda_X/\partial T$ and $\partial \lambda_Y/\partial T$ are the temperature coefficients and $\partial \lambda_X/\partial \varepsilon$ and $\partial \lambda_Y/\partial \varepsilon$ are the transverse deformation coefficients.

Expressions (17) and (18) can be rearranged and written in matrix form in order to calculate the transverse strain and temperature, given the measured wavelength shifts for each polarization band:

$$\begin{vmatrix} \Delta T \\ \Delta \varepsilon \end{vmatrix} = \mathbf{K}^{-1} \begin{vmatrix} \Delta \lambda_X \\ \Delta \lambda_Y \end{vmatrix} \tag{19}$$

where K is a matrix given by

$$K = \begin{vmatrix} \dfrac{\partial \lambda_X}{\partial T}, \dfrac{\partial \lambda_X}{\partial \varepsilon} \\[2ex] \dfrac{\partial \lambda_Y}{\partial T}, \dfrac{\partial \lambda_Y}{\partial \varepsilon} \end{vmatrix} \qquad (20)$$

A simultaneous measurement of transverse strain and temperature can be obtained by determining the coefficients of K, which are determined with previous characterization.

Two examples of simultaneous measurement of these parameters are shown in the table 6 for an IEC fiber and table 7 for a PANDA fiber. The results were obtained using the values of the Bragg wavelength changes for both polarizations bands.

Table 6. Simultaneous measurements of temperature and transverse strain using a FBG written in a IEC HiBi fiber. The set values were determined by the experimental system equipment. Direction of applied load: 0°. [19]

Set values	12 °C	23 °C	31 °C	46 °C
61 με	11.8 °C	24.6 °C	31.5 °C	45.5 °C
	65 με	71 με	70 με	75 με
76 με	12.0 °C	25.4 °C	32.6 °C	46.3 °C
	69 με	79 με	73 με	81 με
91 με	12.7 °C	26.8 °C	34.0 °C	48.0 °C
	76 με	83 με	78 με	79 με

Table 7. Simultaneous measurements of temperature and transverse strain using a FBG in written in a PANDA HiBi fiber. The set values were determined by the experimental system equipment. Direction of applied load: 90°

Set values	7 °C	21 °C	22 °C	40 °C	53 °C
11 με	7 °C	20 °C	23 °C	39 °C	51 °C
	12 με	12 με	9 με	11 με	10 με
21 με	7 °C	19 °C	22 °C	38 °C	51 °C
	21 με	20 με	17 με	16 με	18 με
31 με	6 °C	18 °C	22 °C	38 °C	53 °C
	32 με	31 με	31 με	22 με	24 με
41 με	5 °C	18 °C	21 °C	37 °C	55 °C
	39 με	44 με	44 με	37 με	39 με
51 με	5 °C	18 °C	21 °C	37 °C	55 °C
	43 με	42 με	48 με	41 με	43 με

5.2. Simultaneous Measurement of Transverse Strain and Longitudinal Strain

For the measurement of the longitudinal ($\Delta\varepsilon_Z$) and transverse ($\Delta\varepsilon_X$ or $\Delta\varepsilon_Y$) strain, the equations can also be written in matrix form, given the measured wavelength shifts for each polarization band:

$$\begin{vmatrix} \Delta\varepsilon_Z \\ \Delta\varepsilon_{X,Y} \end{vmatrix} = K^{-1} \begin{vmatrix} \Delta\lambda_X \\ \Delta\lambda_Y \end{vmatrix} \tag{21}$$

where K is now given by:

$$K = \begin{vmatrix} \dfrac{\partial\lambda_X}{\partial\varepsilon_Z}, \dfrac{\partial\lambda_X}{\partial\varepsilon_{X,Y}} \\[3mm] \dfrac{\partial\lambda_Y}{\partial\varepsilon_Z}, \dfrac{\partial\lambda_Y}{\partial\varepsilon_{X,Y}} \end{vmatrix} \tag{22}$$

Table 8 shows an example of simultaneous measurements of longitudinal and transversal strain obtained using the wavelength changes of both polarizations bands.

Table 8. Simultaneous measurements of longitudinal and transverse strain using an FBG written in a IEC HiBi fiber. The set values were determined by the experimental system equipment. Direction of applied load: 90°

Set values	0 με	9 με	14 με
83 με	1 με	8 με	13 με
	64 με	66 με	68 με
167 με	1 με	8 με	13 με
	160 με	161 με	154 με
251 με	0 με	7 με	12 με
	240 με	246 με	244 με
335 με	1 με	8 με	13 με
	320 με	316 με	313 με

5.3. Simultaneous Measurement of Transverse Strain, Longitudinal Strain and Temperature

Two superimposed Bragg gratings can be written in high birefringence fiber optics to measure simultaneously temperature, transverse and longitudinal strain.

This section demonstrates the use of a pair of Bragg gratings written in high birefringence fiber optics to measure, simultaneously, three physical parameters [31]. The Bragg gratings are superimposed in the same position of the fiber optic, in order to behave as a single sensor with reduced dimension.

5.3.1. Superimposed Bragg Gratings

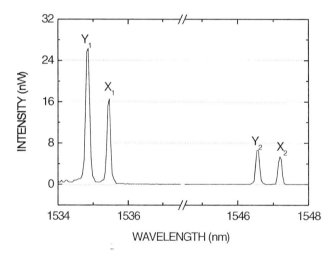

Figure 14. Optical reflection spectrum of two superimposed Bragg gratings written in HiBi IEC fiber [31].

Figure 14 shows an optical reflection spectrum of two gratings recorded at the same fiber position. The two FBG were written with different periods in an IEC HiBi optical fiber with 125 μm diameter. The figure shows the polarization bands (Y-polarization and X-polarization) of each pair. Their relative intensity is not the same as the optical source was not flat along the full wavelength range.

The superimposed HiBi FBGs were characterized by longitudinal, transversal strain and temperature. The measurements of transversal load were made with the fiber oriented with the fast or slow birefringence axis in the direction of the applied load.

Figure 15 shows the dependence of the peak position of each reflection band against the transversal strain applied to the sample (load applied along the Y-axis direction). The best-fitted lines are not parallel; their slopes are different depending on the polarization band. This asymmetric behavior can be used to distinguish the effects of longitudinal and transversal strain acting upon the grating pair.

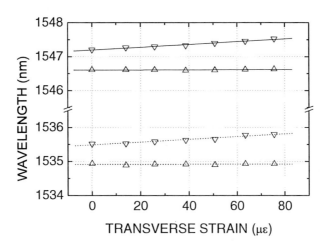

Figure 15. Dependence of the peak wavelength on transverse strain for the reflection bands [X (∇) and Y ($\triangle$)] of the two superimposed FBGs written in an IEC HiBi fiber. Direction of applied load: 90°. The lines represent the linear best fit to the experimental data [31].

The behavior of the reflection bands, when the sensor is under longitudinal strain, is the same for both gratings. The temperature dependence of the reflection bands of the both FBGs has also approximately the same behavior, which is an increase in the wavelength with an increase of temperature.

Table 9. Slopes of temperature, longitudinal and transverse strain for the two superimposed FBGs in an IEC HiBi fiber. Direction of applied transverse load: Y-axis [31]

Slopes	Polarization bands			
	Y_1	X_1	Y_2	X_2
$\partial\lambda/\partial T$ (pm/°C)	8.4	7.8	7.8	7.5
$\partial\lambda/\partial\varepsilon_Y$ (pm/µε)	0.08	4.02	0.19	4.11
$\partial\lambda/\partial\varepsilon_Z$ (pm/µε)	1.3	1.39	1.39	1.36

The corresponding slopes of temperature, longitudinal and transversal strain for both polarization bands, for the best-fitted lines of superposing FBGs in IEC HiBi fiber, are given in table 9.

5.3.2. Simultaneous Measurements

The change in the Bragg wavelength of the reflection spectrum of the both FBGs, due to a temperature change ΔT, a transverse strain ($\Delta \varepsilon_X$ or $\Delta \varepsilon_Y$) and longitudinal strain $\Delta \varepsilon_Z$, for each polarization, is given by

$$\Delta \lambda_{X1} = \frac{\partial \lambda_{X1}}{\partial T} \Delta T + \frac{\partial \lambda_{X1}}{\partial \varepsilon_{X,Y}} \Delta \varepsilon_{X,Y} + \frac{\partial \lambda_{X1}}{\partial \varepsilon_Z} \Delta \varepsilon_Z \qquad (23)$$

$$\Delta \lambda_{Y1} = \frac{\partial \lambda_{Y1}}{\partial T} \Delta T + \frac{\partial \lambda_{Y1}}{\partial \varepsilon_{X,Y}} \Delta \varepsilon_{X,Y} + \frac{\partial \lambda_{Y1}}{\partial \varepsilon_Z} \Delta \varepsilon_Z \qquad (24)$$

$$\Delta \lambda_{X2} = \frac{\partial \lambda_{X2}}{\partial T} \Delta T + \frac{\partial \lambda_{X2}}{\partial \varepsilon_{X,Y}} \Delta \varepsilon_{X,Y} + \frac{\partial \lambda_{X2}}{\partial \varepsilon_Z} \Delta \varepsilon_Z \qquad (25)$$

$$\Delta \lambda_{Y2} = \frac{\partial \lambda_{Y2}}{\partial T} \Delta T + \frac{\partial \lambda_{Y2}}{\partial \varepsilon_{X,Y}} \Delta \varepsilon_{X,Y} + \frac{\partial \lambda_{Y2}}{\partial \varepsilon_Z} \Delta \varepsilon_Z \qquad (26)$$

where $\partial \lambda_{X1}/\partial T$, $\partial \lambda_{X2}/\partial T$, $\partial \lambda_{Y1}/\partial T$ and $\partial \lambda_{Y2}/\partial T$ are the temperature coefficients, $\partial \lambda_{X1}/\partial \varepsilon_{X,Y}$, $\partial \lambda_{X2}/\partial \varepsilon_{X,Y}$, $\partial \lambda_{Y1}/\partial \varepsilon_{X,Y}$ and $\partial \lambda_{Y2}/\partial \varepsilon_{X,Y}$ are the transversal deformation coefficients, and $\partial \lambda_{X1}/\partial \varepsilon_Z$, $\partial \lambda_{X2}/\partial \varepsilon_Z$, $\partial \lambda_{Y1}/\partial \varepsilon_Z$ and $\partial \lambda_{Y2}/\partial \varepsilon_Z$ are the longitudinal deformation coefficients.

Equations (23) to (26) can be rearranged and written in matrix form, in order to calculate the transverse, longitudinal strain and temperature, given the measured wavelength shifts for each polarization band. In this way, the calculation of the three parameters being measured can be made using the following (the choice of reflection bands was arbitrary):

$$\begin{vmatrix} \Delta T \\ \Delta \varepsilon_{X,Y} \\ \Delta \varepsilon_Z \end{vmatrix} = K^{-1} \begin{vmatrix} \Delta \lambda_{Y1} \\ \Delta \lambda_{X1} \\ \Delta \lambda_{Y2} \end{vmatrix} \qquad (27)$$

where K is assembled from the several sensitivities for temperature and deformation:

$$K = \begin{vmatrix} \dfrac{\partial \lambda_{Y1}}{\partial T} & \dfrac{\partial \lambda_{Y1}}{\partial \varepsilon_{X,Y}} & \dfrac{\partial \lambda_{Y1}}{\partial \varepsilon_Z} \\[2mm] \dfrac{\partial \lambda_{X1}}{\partial T} & \dfrac{\partial \lambda_{X1}}{\partial \varepsilon_{X,Y}} & \dfrac{\partial \lambda_{X1}}{\partial \varepsilon_Z} \\[2mm] \dfrac{\partial \lambda_{Y2}}{\partial T} & \dfrac{\partial \lambda_{Y2}}{\partial \varepsilon_{X,Y}} & \dfrac{\partial \lambda_{Y2}}{\partial \varepsilon_Z} \end{vmatrix} \qquad (28)$$

After a previous characterization, in order to obtain K, the temperature, longitudinal and transversal strain components can be simultaneously measured. Some of the obtained results with the grating pair described above are given in table 10.

Table 10. Simultaneous measurements of temperature, transverse and longitudinal strain using two superimposed FBGs in IEC HiBi fiber. The set values were determined by the experimental system equipment [31].

Set values	167 $\mu\varepsilon$		251 $\mu\varepsilon$	
	15 °C	**45 °C**	**15 °C**	**45 °C**
	12 °C	42 °C	12 °C	37 °C
12 $\mu\varepsilon$	13 $\mu\varepsilon$	10 $\mu\varepsilon$	16 $\mu\varepsilon$	10 $\mu\varepsilon$
	117 $\mu\varepsilon$	99 $\mu\varepsilon$	228 $\mu\varepsilon$	177 $\mu\varepsilon$
	16 °C	33 °C	16 °C	43 °C
22 $\mu\varepsilon$	18 $\mu\varepsilon$	16 $\mu\varepsilon$	16 $\mu\varepsilon$	24 $\mu\varepsilon$
	141 $\mu\varepsilon$	132 $\mu\varepsilon$	252 $\mu\varepsilon$	187 $\mu\varepsilon$
	16 °C	36 °C	18 °C	43 °C
32 $\mu\varepsilon$	32 $\mu\varepsilon$	29 $\mu\varepsilon$	21 $\mu\varepsilon$	32 $\mu\varepsilon$
	116 $\mu\varepsilon$	139 $\mu\varepsilon$	236 $\mu\varepsilon$	178 $\mu\varepsilon$

5.4. Bragg Gratings in Reduced Diameter High Birefringence Fiber Optics

Bragg gratings written in reduced diameter high birefringence fiber optics can also be used for multiparameter sensing. Changes in the stress profile of HiBi fibers due to reduced diameter can modify the response of a FBG sensor system to strain or temperature optimizing the simultaneous measurement of those parameters. Chemical etching can be a good tool to reduce the fiber diameter. The changes in the birefringence properties of HiBi fibers as a function of fiber diameter can be analyzed using fiber samples chemically etched in hydrofluoric acid (HF), while the optical spectra of pre-recorded gratings are measured [34].

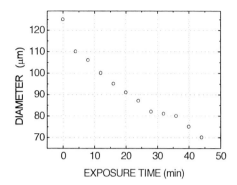

Figure 16. Diameter of an IEC HiBi fiber as a function of the exposure time. HF concentration: 20% [34].

The diameter of the fibers during the etching can be measured by having several samples of the fiber in the acid. The samples are removed successively from the acid, rinsed in distilled water, dried, and then measured under a microscope with a calibrated scale.

The evolution of the diameter, as a result of etching, for an IEC fiber is presented in figure 16. HF acid was diluted to 20 % (parts per volume) in order to reduce the velocity of chemical etching and to increase the sampling points along the process. Figure 17 shows the changes in the transversal section of the IEC fiber, with 125 µm of diameter (left) and after etching (right), with 86 µm of diameter. The internal elliptical cladding can be observed in these photographs. The major axis of the ellipse has approximately 75 µm. The etched IEC fiber shows a higher asymmetry on the borders close to the axes along the major axis of the internal elliptical cladding.

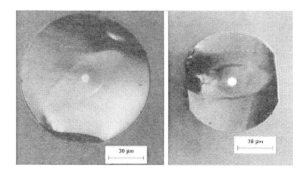

Figure 17. Microphotographs of the transverse section of an IEC HiBi fiber. Left: standard HiBi fiber with 125 µm of diameter. Right: etched HiBi fiber with 86 µm of diameter [34].

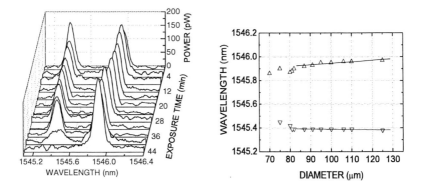

Figure 18. Left: evolution of the reflection bands of a FBG written in an IEC HiBi fiber as a function of the etching time. Right: peak position of the polarized bands (Y-polarized (∇) and X-polarized (Δ))as a function of the fiber diameter. The lines represent the linear best fit for the experimental data. HF concentration: 20 % [34].

Figure 18 (left) illustrates the optical reflection spectra of the FBG in IEC fiber, obtained as a function of HF exposure time. After 36 minutes of exposition time, the optical spectrum had a single band, which means that, the fiber birefringence was almost zero. That is a consequence of the stress release due to the etching.

Figure 18 (right) shows the changes in the peak position of the reflected polarized bands as a function of the IEC fiber diameter. The different slopes for the X and Y polarized bands can be related to asymmetric changes of the internal stress applied by the internal elliptical cladding.

The evolution of the birefringence, as a function of the diameter, can be seen in figure 19.

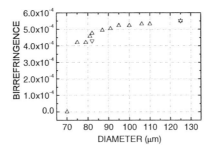

Figure 19. Calculated birefringence of the IEC HiBi fiber as a function of diameter. HF concentration: 40 % (∇) and 20 % (Δ) [34].

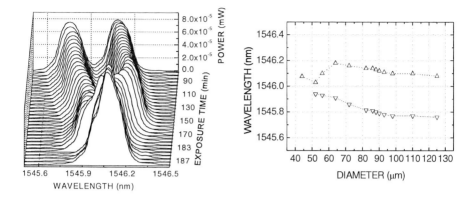

Figure 20. Left: evolution of the polarized bands of a FBG written in a bow tie HiBi fiber as a function of the etching time. Right: peak position of polarized bands (Y-polarized (∇) and X-polarized (Δ)) as a function of the fiber diameter. The lines represent the linear best fit for the experimental data. HF concentration: 20 % [34].

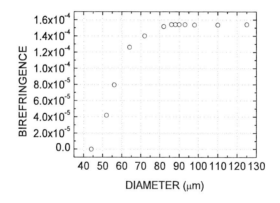

Figure 21. Measured birefringence of bow tie HiBi fiber as a function of diameter. HF concentration: 20 % [34].

A similar characterization can be made to other types of HiBi fibers. For example, figure 20 (left) shows the effect of chemical etching in the optical spectrum of a Bragg grating written in a bow tie fiber. The etching rate is lower and it is possible to observe that the two polarization bands collapse. Initially both bands show a trend to longer wavelengths on their peak position, as the diameter changes from 100 µm to 65 µm (figure 20 (left)). Further etching now causes the X polarized band to shift sharply to shorter wavelengths, until both bands collapse when the diameter reaches approximately 40 μm. This value agrees with the intrinsic stress-applying region dimensions, where the distance between the boundaries of the two internal side-lobes is approximately 37 μm.

Figure 21 shows the birefringence for a bow tie fiber as a function of the diameter. The results show that IEC and bow tie fibers have vanishing birefringence for diameters that are close to the value of the maximum dimension of the stress-applying region.

5.4.1. Reduced Diameter for the Simultaneous Measure of Transverse Strain and Temperature

A FBG in an etched HiBi fiber can be applied as a sensor to simultaneously measure the transverse strain and temperature. Once again, a previous calibration of the different sensitivities must be made. The temperature and transverse strain coefficients for an etched IEC fiber is shown in table 11. It also displays the coefficients for a non-etched bow tie fiber with a similar diameter.

Table 11. Slopes of temperature and transverse strain of a FBG written in etched IEC and non-etched bow tie HiBi fibers [34]

Fiber (diameter)	Temperature		Transversal strain	
	$\partial\lambda_x/\partial T$ (pm/°C)	$\partial\lambda_y/\partial T$ (pm/°C)	$\partial\lambda_x/\partial\varepsilon$ (pm/με)	$\partial\lambda_y/\partial\varepsilon$ (pm/με)
Etched IEC (82 μm)	7.00	6.90	0.7 (X -axis) 3.4 (Y -axis)	2.23(X -axis) 0.1 (Y -axis)
Bow tie (80 μm)	8.02	8.46	0.02 (X -axis) 1.2 (Y -axis)	1.16 (X -axis) 0.3 (Y -axis)

The results of simultaneous transversal strain and temperature measurements obtained with matrix K and the values of $\Delta\lambda_x$ and $\Delta\lambda_y$ of the reflection spectra are displayed in table 12.

The errors obtained using a FBG in normal and reduced diameter HiBi fibers as a sensor are of comparable magnitude, but the dynamic range for strain measurements with the later ones is almost doubled as compared to the former sensors. This fact is important for technological applications where FBG can be tailored to attend a specific measurement range.

Table 12. Simultaneous measurements of temperature and transverse strain using etched FBGs in IEC HiBi fiber (diameter of 82 µm). The set values are determined by the experimental system equipment. Direction of applied load: 90° [34]

Set values	16 °C	26 °C	36 °C	46 °C	56 °C
33 µε	15 °C	28 °C	33 °C	44 °C	56 °C
	37 µε	37 µε	39 µε	42 µε	42 µε
48 µε	17 °C	29 °C	36 °C	47 °C	56 °C
	54 µε	57 µε	56 µε	54 µε	59 µε
64 µε	17 °C	28 °C	36 °C	46 °C	57 °C
	49 µε	48 µε	48 µε	54 µε	53 µε
79 µε	17 °C	29 °C	36 °C	48 °C	58 °C
	66 µε	80 µε	71 µε	65 µε	73 µε
94 µε	17 °C	29 °C	36 °C	48 °C	58 °C
	80 µε	100 µε	94 µε	91 µε	85 µε

6. APPLICATIONS TO OPTICAL COMMUNICATIONS

All optical processing devices are becoming a key element in the next generation of optical communication systems, since they play a critical role in pulse formatting, spectral shaping and optimized all-optical routing and switching. These devices don't have the typical bottleneck associated to the optical-electrical-optical conversion and the majority is transparent to modulation format and bit-rate. FBGs are quite interesting for these applications, due to their low insertion loss and due to the avoidance of the decoupling of the signal outside the fiber. Moreover, the production technology is now in a mature state, which enables the design of gratings with custom made transfer functions, crucial for all-optical processing. Some advanced processing can be made if the transfer function is different for the two transversal modes of propagation in the fiber. This can be achieved by a HiBi FBG. One of the devices that take full advantage of the optical processing capabilities of the HiBi FBG is the orthogonal pumps source [35-37], which can be used in all optical wavelength converters [38, 39]. A tunable PMD compensator can also be developed based on the polarization processing properties of these special gratings [40, 41]. Also, a tunable microwave-photonic notch filter that makes use of a time delay element based on tunable HiBi chirped FBG has been demonstrated [42, 43] In addition, the interference due to laser coherence, typical in those micro-wave photonic filters was also reduced due to the polarization properties of the HiBi FBGs.

The following sections describe some example application of HiBi FBG in optics communications.

6.1. Optical Delay Line for PMD Compensation

In a linearly chirped grating, written in a HiBi fiber, each position of the grating will reflect two wavelengths at orthogonal polarizations (figure 22). This means that the group delay of these gratings is a combination of two linear functions, one for each polarization, with the same slope (D_{FBG}) and shifted by $\Delta\lambda_{HB}$:

$$\tau_y(\lambda) = D_{FBG}\lambda + b$$
$$\tau_x(\lambda) = D_{FBG}(\lambda - \Delta\lambda_{HB}) + b \tag{29}$$

where b in (29) is a constant.

Therefore, the relative group delay induced by a linearly chirped FBG written in a HiBi fiber ($\Delta\tau = \tau_x - \tau_y$) is calculated using the following expression

$$\Delta\tau = -D_{FBG}\Delta\lambda$$
$$\approx -2D_{FBG}B\Lambda \tag{30}$$

Expression (30) shows that the dynamic tuning of the induced PMD can be made by adjusting the birefringence of the fiber, which can be done by applying a transversal stress in the fiber, as shown before in this chapter.

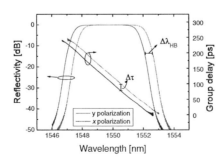

Figure 22. Reflectivity and group delay of a linearly chirped HiBi FBG for both transversal propagation modes [47].

6.1.1. Compensation Using a Linear Chirp

As can also be observed in expression (30), it is also possible to tune the PMD by adjusting the dispersion of the grating. That can be done using different methods [44-46]. One of them is by using thermal gradients to induce a linear chirp to a uniform FBG. Let us consider a uniform HiBi FBG put in a thermal contact with metal substrate. By applying different temperatures to the substrate, different linear temperature gradients will be generated. This gradient will induce a linear chirp to the FBG, due to thermo-optic and photoelastic effects. By changing the temperature gradient, the dispersion will also change, inducing a tunable differential delay line [47]. Figure 23 shows the experimental results of the evolution of $\Delta\tau$ as a function of the applied temperature gradient to a 24 mm uniform HiBi FBG.

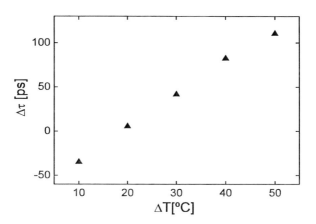

Figure 23. Relative group delay as a function of the applied linear gradient to a uniform HiBi FBG with 24 mm length.

Therefore, the presented device can be included in a PMD compensator as a tunable optical relative group delay line.

6.1.2. Compensation Using a Nonlinear Chirp

Let us now suppose that we have a HiBi FBG with a quadratic chirp. The group delay is now composed by two parabolic functions (one for each polarization) shifted by $\Delta\lambda_{HB}$. If the grating is tuned by temperature or longitudinal stress, the relative induced delay between the orthogonal polarizations, for a specific wavelength will change [40]. Figure 24 shows a simulation of a quadratic chirped FBG, with a length of 25 mm, written in a HiBi fiber with birefringence $B = 5 \times 10^{-4}$. For a tuning of 4.5 nm in the central wavelength, the relative group delay at 1550 nm changed from 41.6 ps to 12.1 ps. In this way, with this method, it is possible to do small corrections in the relative group delay.

The advantages of this method are its tuning simplicity and the flexibility in the operation range. However, the technique needs a FBG with a nonlinear chirp, which is quite complex to produce. It is generally produced with a custom made phase mask with a nonlinear chirp.

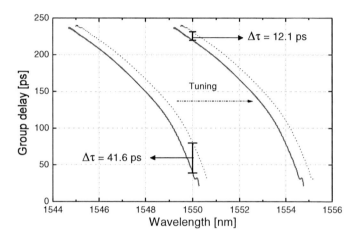

Figure 24. Simulation of the group delay of a HiBi FBG with quadratic chirp. Line: Y polarization; dots: X polarization [47].

6.2. Tunable Multiwavelength Linear Polarized Fiber Lasers

Fiber lasers have different applications in sensors and telecommunications due to their reduced linewidth, power and spectral profile. Like other lasers, fiber lasers need two components: a gain medium and a resonant cavity. For a fiber laser operating around 1550 nm, it is generally based on an optical pump with 980 or 1480 nm of wavelength, an erbium-doped fiber and an optical filter. The gain is obtained from the amplified spontaneous emission due to the optical pump.

Generally, fiber optical lasers based on an optical ring with erbium-doped fiber don't enable the generation of more than one laser line [48,49]. This is a consequence of the fact that erbium is a medium with homogeneous gain at room temperature, resulting in strong mode competition, which induces laser instability. A method was proposed to reduce the homogeneity of the fiber by cooling the fiber to 77 K [50, 51]. However, by obvious reasons, it is not very practical. Other methods used special fibers like the elliptical core fibers [52] or the twincore fibers [53].

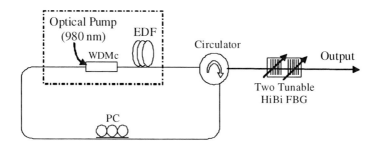

Figure 25. Diagram of a multiwavelength fiber laser based on HiBi FBGs. EDF: Erbium doped fiber; PC: Polarization controller; WDMc: WDM optical coupler;

Another way to reduce the homogeneity of the fiber is to use different laser lines operating at different longitudinal modes. For this kind of implementation, HiBi FBGs can have an important role, since they will reflect two wavelengths at orthogonal polarizations. An implementation method for a tunable laser with up to four laser lines is depicted in figure 25.

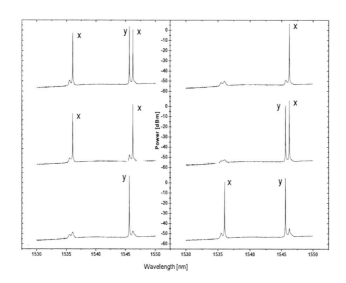

Figure 26. Optical spectra at the output of the fiber laser with different operation modes. The operating laser lines are at a linear polarization (x or y).

The two tunable HiBi FBGs enable the selection of 4 different wavelengths. By tuning the polarization controller (PC) inside the optical cavity, it is possible to select the appropriate laser lines. Figure 26 shows some of the possibilities that can be achieved with just two HiBi FBGs.

One of the advantages of this technique is its ability to generate two laser lines at orthogonal polarizations (see last spectrum of figure 26). Therefore, it can be used as two orthogonal pumps in a polarization insensitive wavelength converter [38].

6.3. Optical Networks Architectures Using HiBi FBG for Performance Improvement

6.3.1. Optical Code Division Multiple Access

Metro optical code division multiple access (OCDMA) networks can benefit from the polarization multiplexing, since two users using codes in the same time-wavelength chip can be given orthogonal polarizations to operate, therefore reducing interference. One of the implementation techniques is the "polarization assisted OCDMA with HiBi FBG" [54]. The technique uses the polarization properties of the HiBi FBG along with a special code generation scheme to improve the performance of OCDMA based networks. The coders are based on HiBi FBGs. To implement the suggested polarization assisted OCDMA, each HiBi FBG will reflect a pair of wavelengths $\lambda_i \lambda_j$, which are consecutive and cross polarized. In the case of the proposed method, a set of three of these HiBi FBGs, spaced by the fiber length needed for achieving the corresponding time chip spacing, results in two subsequent codes. Here, λ_i corresponds to a X polarized reflection and λ_j to a Y polarized one. This allows two consecutive user spreading sequences to share the same encoder. An implementation example is depicted in figure 27.

Each bit of information from users A and B is a wavelength comb which includes at least the wavelengths of the correspondent code (or a standard modulated broadband source can be used). Both bit sequence signals are multiplexed using a polarization beam combiner, thereby ensuring that they enter the encoder at the correct orthogonal polarizations. The encoder is based on three HiBi FBG reflecting the wavelengths λ_3, λ_{11} and λ_{24} for X polarization and λ_4, λ_{12} and λ_{25} for Y polarization. To achieve networking operation many such encoders need to operate simultaneously, and due to the properties of the technique, the number of encoders needed reduces to almost half.

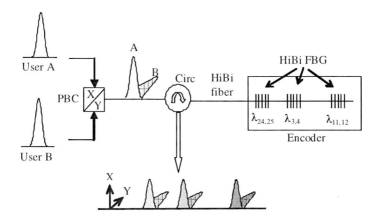

Figure 27. Schematic of the proposed encoder implementation showing the use of the polarization to encode simultaneously two users with different wavelengths at orthogonal polarizations. Legend: PBC: polarization beam combiner; Circ: optical circulator [54] (© 2006 IEEE).

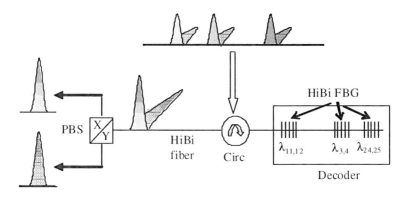

Figure 28. Schematic of the proposed implementation for the decoder based on HiBi FBG. Legend: PBS: polarization beam splitter; Circ: optical circulator [54] (© 2006 IEEE).

The decoder can be imprinted with standard FBG, since each user has its own code. However, for reduced user interference and to reduce the number of decoders needed, it can be based on HiBi FBG like the one exemplified in figure 28.

The decoding process is similar to the encoding, where the HiBi FBG correlates two codes simultaneously. Afterwards a polarization beam splitter is used to separate both users. If no polarization maintaining fiber is used in the transmission link between the encoder and the decoder, the former must be preceded by a polarization rotator to ensure correct polarization coupling to the receiver. The polarization rotator can be automatically controlled by the receiver of one of the users using simple electronics. Even if no alignment of the polarization is made between non adjacent users, on average, only half the power will induce interference since the decoder will process only one of the two available polarizations. In the same way, the sensitivity to heterodyne crosstalk is also reduced since the power of the adjacent user, generated by the same encoder, is orthogonally polarized. In opposition to other coding/decoding techniques, like the ones based on arrayed waveguide gratings and optical delay lines, the proposed coder/decoders are quite compact, simple to use and have low insertion losses. On the other hand, since the gratings can have a length down to 1-2 mm and still have a high reflectivity, the time slots can be as low as a few picoseconds which can be considered enough for the majority of applications.

6.3.2. Radio over Fiber

In radio over fiber systems (RoF), using the same central station to transmit to different local stations, one can use frequency interleaving to improve the bandwidth efficiency, exploiting the unused band between the carrier and data when high modulation frequencies are used with single side band (SSB) format. However, frequency interleaving also increases the bit error rate (BER), due to the interference of adjacent carriers. This drawback can be minimized if polarization multiplexing is used, i.e., the carriers and data are at orthogonal polarizations (figure 29).

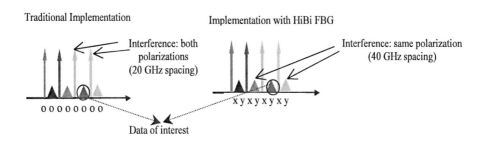

Figure 29. Diagram of the concept of interleaving using polarization multiplexing between carriers and data.

The implementation of this concept can be made using a HiBi FBG filter at the transmission which creates the SSB format and, at the same time, selects one polarization for the carrier and the orthogonal one for the data. At the local station another HiBi FBG removes the selected channel with reduced interference, since the interference will only be made by the data of the adjacent channels, which are at higher wavelength spacing and with lower power, relatively to the adjacent carriers. This technique has an impact on the overall performance of the system since the bandwidth efficiency can be improved without increasing the BER [55].

CONCLUSION

This chapter described some of the characteristics and functionalities associated with HiBi FBG. Their anisotropic behavior, relative to stress and/or strain, make them well suited for multiparameter sensors, including temperature, transversal strain and longitudinal strain. Moreover, their polarization processing capabilities also give them an interesting potential for different applications in optics communications. These applications include the development of new devices, like multiwavelength fiber lasers or in the optimization of certain architectures, like OCDMA. Some of the applications for sensing and optical communications were described but many more are yet to come.

REFERENCES

[1]   Kapron, F. P.; Keck, D. B. *Appl. Phys. Lett.* 1970, vol.17, 423-425.
[2]   Yamada, Y.; Nakagawa, S. I.; Kurosawa, Y.; Kawazawa, T.; Taga, H.; Goto, K. *Elect. Lett.* 2002, vol. 38, 328-330.
[3]   Hill, K. O.; Fufii Y.; Johnson D. C.; Kawasaki, B. S. *Appl. Phys. Lett.* 1978, vol. 32, 647-649.
[4]   Mora, J.; Díez, A.; Cruz, J. L.; Andrés, M. V. *IEEE Photon. Technol. Lett.* 2000, vol. 12, 1680-1682.
[5]   Guan, B.; Tam, H.; Liu, S. *IEEE Photon. Technol. Lett.* 2004, vol. 16, 224-226.
[6]   Tjin, S. C.; Suresh, R.; Ngo, N. Q. J. *Lightwave Technol.* 2004, vol. 22, 1728–1733.
[7]   Liu, Y.; Chiang, K. S.; Chu, P. L. *IEEE Photon. Technol. Lett.* 2005, vol. 17, 450-452.

[8] Kashyap, R. *Fiber Bragg gratings*, Publisher: Academic Press, San Diego, CA, 1999.

[9] Othonos, A.; Kalli, K. *Fiber Bragg gratings*, Publisher: Artech House, Norwood, NA, 1999.

[10] Higuera, J. M. L. *Handbook of Optical Fiber Sensing Technology*; Publisher: John Wiley & Sons, New York, NY, 2002.

[11] Erdogan, T. J. *Lightwave. Technol.* 1997, vol. 15, 1277-1294.

[12] Lam, D. K. W.; Garside, B. K. *Appl. Optics.* 1981, vol. 20, 440-456.

[13] Yariv, A. J. *Quantum Electron.* 1973, vol. 9, 919-933.

[14] Yariv, A.; Nakamura, M. J. *Quantum Electron.* 1977, vol. 13, 233-253.

[15] Dyott, R. B. *Elliptical Fiber Waveguides*; Publisher: Artech House, Boston, MA, 1995.

[16] Rashleigh, S. C. J. *Lightwave. Technol.* 1983, vol. LT-1, 312-330.

[17] Kersey, A. D.; Davis, M. A.; Patrick, H. J.; LeBlanc, M.; Koo, K. P.; Askins, C. G.; Putnam, M. A.; Friebele, E. J. J. *Lightwave Technol.* 1997, vol. 15, 1442-1463.

[18] Othonos, A. *Rev. Sci. Instrum.* 1997., vol. 68, 4309-4341,

[19] Abe, I.; Kalinowski, H. J.; Nogueira, R. N.; Pinto, J. L.; Frazão, O. *IEE P-Circ. Dev. Syst.* 2003, vol. 12, 495-500.

[20] Frazão, O.; Pereira, D. A.; Santos, J. L.; Araújo, F. M.; Ferreira, L. A. *Proc. SPIE* 2005, vol.5855, 765-758.

[21] Chen, G; Liu, L; Jia, H; Yu, J; Xu, L.; Wang, W. *IEEE Photon. Technol. Lett.* 2004, vol. 16, 221-223.

[22] Urbanczyk, W.; Chmielewska, E.; Bock, W. *Meas. Sci. Technol.* 2001, vol. 12, 800-804.

[23] Ferreira, L. A.; Araújo, F. M.; Santos, J. L.; Farahi, F. *Opt. Eng.* 2000, vol. 39, 2226-2234.

[24] Sudo, M.; Nakai, M.; Himeno, K.; Suzaki, S.; Wada, A.; Yamauchi, R. *Proc. 12 th Int. Conf. Optical Fiber Sensors* 1997, vol. 16, 170-173.

[25] Chehura, E.; Ye, C. C.; Staines, E. S.; James, S. W.; Tatam, R. P. *Smart Mater. Struct.* 2004, vol.13, 888-895.

[26] Ye, C.C.; Staines, S. E.; James, S. W.; Tatam, R. P. *Meas. Sci. technol.* 2002, vol. 13, 1446-1449.

[27] Bosia, F.; Giaccari, P.; Facchini, M.; Botsis, J.; Limberger, H.; Salathé, R. *Proc. SPIE* 2002, vol. 4694, 175-186.

[28] Lawrence, C. M.; Nelson, D. V.; Udd, E.; Bennett, T. *Exp. Mech.* 1999, vol. 39,202-209.

[29] Lawrence, C. M.; Nelson, D. V.; Udd, E. *Proc. SPIE* 1997, vol. 3042, 218-228.

[30] Caucheteur, C.; Ottevaere, H.; Nasilowski, T.; Chah, K.; Statkiewicz, G.; Urbanczyk, W.; Berghmans, F.; Thienpont, H.; Mégret, P. *Proc. SPIE* 2005, vol. 5952, 1-10.

[31] Abe, I.; Kalinowski, H. J.; Frazão, O.; Santos, J. L.; Nogueira, R. N.; Pinto, J. L. *Meas. Sci. Technol.* 2004, vol. 15, 1453-1457.

[32] Udd, E.; Schulz, W. L.; Seim, J. *Proc. SPIE* 1999, vol. 3538, 206-214.

[33] Udd, E.; Nelson, D., Lawrence, C.; Ferguson, B. *Proc. SPIE* 1996, vol. 2718,104-107.

[34] Abe I.; Frazão O.; Schiller M. W.; Nogueira R. N.; Kalinowski H. J.; Pinto, J. L. *Meas. Sci. Technol.* 2006, vol.17, 1477–1484.

[35] Nogueira, R. N.; Teixeira, A. L. J.; André, P. S.; Rocha, J. F.; Pinto, J. L. *Proc. Conf. Lasers Electro-Optics* 2003, 545.

[36] André, P. S.; Nogueira, R. N.; Teixeira, A. L.; Lima, M. J. N.; Rocha, R. F.; Pinto, J. L. *Laser Physics Lett.* 2004, vol. 1, 1 - 4.

[37] Zhao, C. L.; Yang, X.; Ng, J. H.; Dong, X.; Guo, X.; Wang, X.; Zhou, X.; Lu, C. *Microw. Opt. Technol. Lett.* 2004, vol. 41, 73 – 75.

[38] Nogueira, R. N.; Teixeira, A. L. J.; André, P. S; Rocha, J. F.; Pinto, J. L. *Opt. Commun.* 2006, vol. 262/1, 38-40.

[39] Nogueira, R. N.; Teixeira, A. L. J.; Pinto, J. L.; Rocha, R. F. *IEE Elect. Lett.* 2004, vol. 40, 616–617.

[40] Lee, S.; Khosravant, R.; Peng, J.; Grubsky, V.; Starodubov, D. S.; Willner, A. E.; Feinberg, J., *IEEE Photon. Technol. Lett.* 1999, vol. 11, 1277–1279.

[41] Willner, A. E.; Feng, K.-M.; Cai, J.; Lee, S.; Peng, J.; Sun, H. *IEEE J. Select. Topics Quantum Electron.* 1999, vol. 5, 1298-1311.

[42] Yi, X.; Lu, C.; Yang, X.; Zhong, W.-D.; Wei, F.; Ding, L.; Wang, Y. *IEEE Photon. Technol. Lett.* 2003, vol. 15, 754-756.

[43] Zhang, W.; Williams, J. A. R.; Bennion, I. *IEEE Photon. Technol. Lett.* 2001, vol. 13, 523-525.

[44] Eggleton, B. J.; Mikkelsen, B.; Raybon, G.; Ahuja, Rogers, A.; J. A.; Westbrook, P. S.; Nielsen, T. N.; Stulz, S.; Dreyer, K. *IEEE Photon. Technol. Lett.* 2000, vol. 12, 1022-1024.

[45] Mora, J.; Ortega, B.; Andrés, M. V.; Capmany, J.; Cruz, J. L.; Pastor, D.; Sales, S. *IEEE Photon. Technol. Lett.* 2003, vol. 15, 951-953.

[46] Lauzon, J.; Thibault, S.; Martin J.; Ouellettet, F. *Opt. Lett.* 1994, vol. 19, 2027-2029.

[47] Nogueira, R. N.; Pinto, J. L.; Rocha, J. F. *Microw. Opt. Tech. Lett.* 2006, vol. 48 , 2357-2359.

[48] Gloag, A. J.; Langford, N.; Bennion I.; Zhang, L. *Opt. Commun.* 1996, vol. 123, 553-557.
[49] Inaba, H.; Akimoto, Y.; Tamura, K.; Yoshida, E.; Komukai T.; Nakazawa, M. *Opt. Commun.* 2000, vol. 180, 121–125.
[50] Yamashita S.; Hotate, K. *Electron. Lett.* 1996, vol. 32, 1298-1299.
[51] Wei, D.; Li, T.; Zhao Y.; Jian, S. *Opt. Lett.* 2000, vol. 25, 1150-1152.
[52] Das G.; Lit, J. W. Y. *IEEE Photon. Technol. Lett.* 2002, vol. 14, 606-608.
[53] Graydon, O.; Loh, W. H.; Laming, R. I.; Dong, L. *IEEE Photon. Technol. Lett.* 1996, vol. 8, 63-65.
[54] Nogueira, R. N.; Teixeira, A. L. J.; Pinto, J. L.; Rocha, J. F. *IEEE Photon. Technol. Lett.* 2006, vol. 18, 841 – 843.
[55] Teixeira, A. T.; Nogueira, R. N.; André, P. S.; Lima, M. J. N.; Rocha, J. F., *Electron. Lett.* 2005, vol. 41, 30 - 32.

In: New Developments in Optics Research   ISBN: 978-1-60324-505-7
Editor: Matthew P. Germanno   © 2012 Nova Science Publishers, Inc

Chapter 5

OPTICS RESEARCH APPLIED TO THE TURIN SHROUD: PAST PRESENT AND FUTURE

G. Fanti[*] *and R. Basso*

Dept. of Mechanical Engineering, University of Padua, Padua, Italy

ABSTRACT

The aim of this work is to summarize the important contribution furnished by Optics in the studies of the TS (Turin Shroud) made in more than a century of researches and tests.

In the first section the very peculiar characteristics of the TS are presented and discussed showing that up to now the double body image impressed on it is not yet reproducible. Some body image formation hypotheses are presented and they will be tested by also means of optical means successively discussed.

The second and third sections respectively present the most important optical researches done in the past and in the present in reference to the Relic and the fourth section discusses some possible optical studies that should be done on the TS to unveil some obscure points. It is also accounted for the conservation problems that are very important for the body image impressed in the linen Cloth in a very particular way.

In any case the complexity of the studies related to the TS implies that the future analyses will not be let to a limited group as it was done in 2002,

---

[*] Corresponding author: Dept. of Mechanical Engineering, University of Padua, Via Venezia 1, 35131 Padua Italy. Phone: +39 049 8276804, e-mail: giulio.fanti@unipd.it, web: www.dim.unipd.it/fanti

but to a very wide commission composed of experts in each one of the many disciplines involved in the studies.

1. INTRODUCTION

1.1. What Is the Turin Shroud

The word "shroud" corresponds to the Italian "Sindone" that derives from the ancient Greek and it means "burial garment in which a corpse is wrapped". The Turin Shroud (TS) is 4.4 m long and 1.1 m wide, on which the complete front and dorsal images of the body of a man are indelibly impressed (figure 1.1). The TS is believed by many to be the burial cloth in which Christ was wrapped before being placed in a tomb in Palestine about 2000 years ago and the Science has not demonstrated the contrary. It is the most important Relic of Christianity and, of all religious relics, it has generated the greatest controversy. From a scientific point of view, this Relic is still unexplainable because up to now the body image has not been reproduced in all its details.

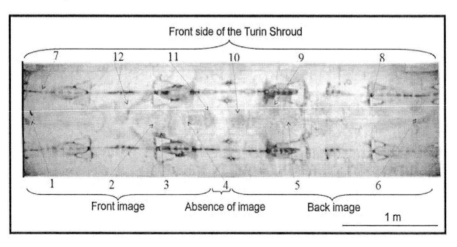

Figure 1.1. Body image and marks visible on TS: 1. Wound in right foot. 2. Marks of water. 3. Wound in side. 4. Folds in cloth. 5. Marks of scourging. 6. Heel and sole of right foot. 7. Carbonized lines in cloth, due to fire of 1532. 8. Mending done by Chambéry nuns after fire of 1532. 9. Bruises due to transport of patibulum. 10. Wounds on head, due to crown of thorns. 11. Wound on forehead. 12. Wound in left wrist (Fanti G. & Maggiolo, 2004).

Even if various probabilistic studies (De Gail, 1972; Fanti & Marinelli, 2001), analysing up to 100 statements formulated for and against the authenticity of the TS, show that it is the burial sheet of Jesus Christ of Nazareth, with a probability of 100% and negligible uncertainty, from a strictly scientific point of view no sure proofs of its authenticity are still available.

According to scientific analysis of the TS in 1978 by the STURP (Shroud of TUrin Research Project) (Jackson J.P. et al., 1984; Jumper et al., 1984; Adler, 1996), it was concluded that the body image on the TS cannot be explained scientifically. One attempt at explanation states that the image formed as if it were caused by exposure to a short-lived but intense source of energy coming from the body enveloped in the shroud itself (Fanti & Maggiolo, 2004).

The TS cloth is hand-woven and each thread (non-constant diameter of about 0.25 mm) is composed of 70-120 linen fibres (Fanti & Moroni, 2002); the cloth has a herringbone 3/1 warp. Although not all scientists are unanimous, it has been demonstrated (Jumper et al., 1984; Adler, 1996) that the linen sheet wrapped the corpse of a man who had been scourged, crowned with thorns, crucified with nails, and stabbed in the side with a lance. Also impressed on the cloth are many other marks due to blood, fire, water and folding, which have greatly damaged the double body image. Of greatest interest to forensic pathologists are the wounds, which appear to be unfakeable (figure 1.1).

In 1988, the TS was radiocarbon-dated to 1260-1390 A.D. (Damon), but the result is questionable. Some researchers have shown that the 1988 sample is not representative of the TS, because its characteristics differ from the main part of the TS (Rogers, 2005). In any case, as science can not reproduce all the characteristics of the body image, the validity of a measurement method which depends on ambient conditions to an incompletely known object must therefore be doubted.

In 2004 (Fanti & Maggiolo, 2004) detected the presence of an image on the back of the TS verifying a researcher's postulate: Jackson (Jackson, 1990) hypothesized a mechanism of formation of an image as the result of a burst of energy from inside the body, which had mechanically become transparent, and he also predicted that an image would be found on the back surface of the TS, but only corresponding to the front image (because this part of sheet would have gone through the body due to the force of gravity, while the part of sheet corresponding to the back image would have remained stationary on the tombstone).

According to statements by STURP scientists (Schwalbe & Rogers, 1982), the images are extremely superficial, in the sense that only the most external fibers of the linen threads are involved. Therefore the body image of the back side of the TS should only be considered a phenomenon of image formation involving the two most superficial parts of the sheet, letting the linen fibesr inside the fabric not be changed. Many energy sources, such as thermal energy, do not satisfy this requirement. A mechanism of formation which does satisfy this condition is a corona discharge (Fanti, Lattarulo & Scheuermann, 2005), explaining the creation of a double superficial image from both theoretical (Lattarulo, 1998) and experimental points of view (De Liso, 2002).

In one statement, Walsh (Walsh, 1963) observed: "The Shroud of Turin is either the most awesome and instructive relic of Jesus Christ in existence ... or it is one of the most ingenious, most unbelievably clever, products of the human mind and hand on record. It is one or the other; there is no middle ground." From a scientific point of view, it would be very interesting to understand how a corpse could have generated such a peculiar image even now not reproducible in all its characteristics. From a religious point of view, it is important to understand what the TS is, because, if it is authentic, it witnessed the event of the burial and Resurrection of Jesus Christ.

The present work contributes to a thorough examination of the optical characteristics of the body image and of the optical techniques that can be applied, with the aim of being a starting point for future researches that could explain what possible source of energy could have caused the image to form on the linen.

1.2. Brief History

The "Shroud of Christ" first appeared in Europe in 1353 at Lirey, France. In 1203, a soldier camping outside Constantinople with the Crusaders noted that a church there exhibited every Friday the cloth in which Christ had been buried, with the figure of his body impressed on it. It is probable that this cloth and the TS are the same also because the face of Christ minted in the Byzantine coins after the VII century BC is very similar to the TS face. Some clues indicate that the TS was in Palestine in the first century BC.

In 1532, a fire damaged the TS while it was conserved at Chambéry in France. The Chambéry nuns later restored the TS by sewing some patches of cloth on the front of it and the so called "Holland cloth" on the back of it. Figure 1.1 shows the signs of Chambéry fire. The TS was property of the Savoy from March, 22, 1453, date of the acquisition of the TS from Anna of Lusignano, wife of the duke Ludovico of Savoy, to March, 18, 1983, date of the gift to the Pope from Umberto II of Savoy. From 1983 the TS is property of the Vatican.

Up to the XIX century AD scientific interest in the TS was present but limited because of the scarceness of direct analyses and the lack of photographs. The interest greatly developed after 1898, when S. Pia photographed the TS and spread in the scientific world photos for independent studies. The interest also increased when S. Pia noticed that the negative image on the TS looked like a photographic positive. In 1931, G. Enrie photographed the TS at higher resolution using an orthochromatic plate.

The most important scientific analysis of the TS was done in 1978 by the STURP (Shroud of TUrin Research Project) (Jackson, 1984, Jumper et al., Schwalbe) that reached the following conclusion: the body image on the TS cannot be explained scientifically, and the only attempt at explanation consists of stating that the image formed as if it were caused by exposure to a short-lived but intense source of energy coming from the body wrapped by the TS itself.

Images of the back of the TS corresponding to the front and back images were acquired in 2000 (Archdiocese of Turin, 2000) and 2002 (Ghiberti, 2002). During the 2002 intervention, all the patches which had been sewn on by the Chambéry nuns were unstitched, revealing the holes in the fabric left by the fire. Lastly, a reinforcing cloth, which no longer allows direct observation of the back of the TS, was sewn back.

1.3. General Characteristics

The TS (figure 1.1) shows a frontal and a dorsal body image, separated by a space between the two images of the head. The images show an adult male, nude, well-proportioned and muscular, with beard, moustache and long hair, and are compatible with a man 175 ± 2 cm tall enveloped in a sheet (Basso et al., 2000). Due to rigor mortis, which began after his crucifixion, the TS Man was not completely supine but had his head tilted forwards, his knees slightly bent and his feet extended, as a result of being nailed to a cross.

The body image impressed on the TS has many peculiar physical and chemical characteristics which, even now, cannot be reproduced. Some of these are reported in literature (Fanti, 24 authors, 2005) and most of them derive from optical researches that will be presented in this text (the corresponding paragraph that discuss the optical technique that leaded to or helped to reach the result is reported in brackets):

G1. The image is a result of concentrations of yellow to light brown fibers (§2.6g, §2.6h, §2.6i).

G2. There are no signs of cementation among the fibres (§2.6h, §2.6i).

G3. The image color resides only on the topmost fibers at the highest parts of the weave and it resides on the thin impurity layer of outer surfaces of the fibers (§2.6h, §2.6i).

G4. The color of the image-areas has a discontinuous distribution along the yarn of the cloth: striations are evident(§2.6g).

G5. The colored coating cannot be dissolved, bleached, or changed by standard chemical agents, but it can be decolorized by reduction with diimide (§2.6c).

G6. The pyrolysis-mass data showed the presence of polysaccharides of lower stability than cellulose on the surface of linen fibers from the TS (§2.6c).

G7. The body image is chemically due to molecular changes in the polysaccharides, i.e., a conjugated carbonyl structure, in a state of dehydration (§2.6c).

G8. The image formed at a relatively low temperature and no fluorescent pyrolysis products were found in image areas (§2.3, §2.6e).

G9. Side images between the front and back body images, including the region between the two heads, are absent (§2.1, §2.2, §2.3).

G10. All the chemical and microscopic properties of dorsal and ventral image fibers are identical (§2.1, §2.2, §2.3, §2.6, §2.7, §3.1, §3.4, §3.8).

G11. Although anatomical details are generally in close agreement with standard human-body measurements, some measurements made on the Shroud image, such as hands, calves and torso, do not agree with anthropological standards, but they are coherent with the gnomonic distortion due to the cloth enveloping. The frontal body image (195 cm long) is compatible, within an uncertainty of +/-2 cm, with the dorsal image (202 m long) if it is supposed that the TS enveloped a corpse having the head tilted forward, the knees partially bent and the feet stretched forwards and downwards (§2.2, §3.3).

G12. The body image shows no evidences of putrefaction signs and there is no evidence for tissue breakdown (§2.1, §2.2).

G13. Reflectance spectra, chemical tests, laser-microprobe Raman spectra, pyrolysis mass spectrometry, and x-ray fluorescence all show that the image is not painted with any of the expected, historically-documented pigments. Chemical tests showed that there is no protein painting medium or protein-containing coating in image. There are no pigments on the body image in a sufficient quantity to explain the presence of an image (§2.6a, §2.6b, §2.6c, §2.6h, §2.6i).

G14. The very high rigidity of the body is evident on the back image especially in correspondence of the buttocks: the anatomical contours of the back image demonstrate minimal surface flattening (§2.2, §3.3).

G15. A body image is visible in areas of body-sheet non-contact zones, such as those between nose and cheek (§2.2, §2.3).

G16. A body image color is visible on the back surface of the cloth in the same position of some anatomic details as for the body image of the frontal surface of the TS. The hair appears more easily but also other details of face (figure 1.2) and perhaps hands appear by image enhancement. The frontal image, at least in correspondence to the head, is then doubly superficial (§3.5).

G17. The limestone found on the feet contains calcium in the form of aragonite. Similar characteristics were found on samples coming from Ecole Biblique tomb in Jerusalem (§2.6h, §2.6i).

G18. The TS face (figure 1.2) shows a sad but majestic serenity (§2.1, §2.2).

G19. The red stains are of human blood, and were formed on the cloth before the body image was produced, because no image exists under them. There is a class of particles on the TS ranging in color from red to orange that test as blood derived residues. They test positively for the presence of protein, hemin, bilirubin, and albumin; give positive hemochromagen and cyanmethemoglobin responses; after chemical generation display the characteristic fluorescence of porphyrins (§2.6c, §2.6i).

G20. Some human blood stains appear on and outside of the body image (§2.3, §2.6f)

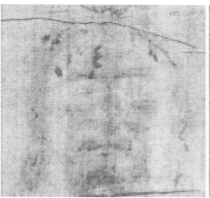

Figure 1.2. Image of face of the TS Man as it appears to observers. On the right mirror image with inverted luminance levels.

1.4. Optical Characteristics

The optical characteristics of the body image are listed below, but it is to evidence that body image is very faint: reflected optical densities are typically less than 0.1 in the visible range and the image shows no evidence of image saturation. It is therefore not easy that a naked eye is able to discriminate all the details codified in it and digital enhancements of photographs are frequently required. Furthermore the photographs of the body images acquired in different ambient conditions such as incident or transmitted, visible or UV or IR light show different details that are important in the understanding of this image that still challenges the Science.

The low contrast of the body image on the background requires a relatively high image enhancement, but this enhancement causes a noise increment that tends to mask much information contained in the photographs; therefore many special kinds of optical cleaning and restoration are required.

Some optical characteristics are the following and most of them derive from optical researches presented in this text (the corresponding paragraph that discuss the technique that leaded to the result is reported in brackets):

O1. If seen in visible light, he body image has the normal tones of light and dark reversed with respect to a photograph, such that parts nearer to the cloth are darker but no reference can be made to the real color of the anatomical details (hair for example); in a simplistic way it can be stated that the body image appears as if it were a **negative** (§2.1).

O2. The luminance distribution of both the frontal and dorsal images has been correlated to the clearances between a three-dimensional surface of the human body and a covering cloth and this luminance distribution is independent of implied body surface composition (e.g., skin, hair, etc.): **3D information** may consequently be codified (§2.5, §2.6j).

O3. The luminance distribution of the body image is consistent with a highly **directional** radiation source, probably normal to the skin (§3.11).

O4. The body image has a resolution of **4.9±0.5 mm at the 5% MTF value** (for example the lips are clearly visible); the resolution of the bloodstains instead is at least ten times better (§3.6).

O5. In the positive photograph of Durante (Archdiocese of Turin, 2000) (perhaps the most resolute color photograph ever done), the **luminance levels** of the front and back body images (face excluded) are **compatible** within an uncertainty of 5% (the front image is generally darker than the dorsal one) and this means that the back image was not influenced by body weight; instead, the luminance level of the face is about 20% higher than that of the mean of the whole body. The hair on the frontal image show high luminance levels relatively to the face: for example the left hair is darker than the cheeks (§3.4).

O6. The body **image does not fluoresce** in the visible under ultraviolet illumination and the non-image area fluoresces with a maximum at about 435 nanometers; a redder fluorescence can be observed only around the burn holes from the 1532 fire; the cloth does not show any phosphorescence (§2.3, §2.6e).

O7. An emission body **image** (positive) is clearly **visible in the 8-14 micrometers IR** range, but IR emission of the image in the 3-5 micrometer range was below the instrument sensitivity (§2.6d).

1.5. Body Image Formation Hypotheses

Both the general and the optical characteristics of the TS are very particular and impossible for the moment to be reproduced all together. This "provocation to the intelligence" as the TS body image was defined by John Paul II, has tickled the minds of many researches who proposed different hypotheses to explain its formation. These hypotheses frequently supported by interesting experimental results both in favour of authenticity, and vice versa have nevertheless been able to reproduce all the details detected on the TS (Fanti, 24 authors, 2005). Examples are:

- The body image is due to an energy source coming from the enveloped man, perhaps during the Resurrection: this source may be of protonic, electronic, ultraviolet (UV) or other type (Scheuermann, 1983; Jackson, 1984; Moran & Fanti, 2002).

- The image, which originated by direct contact of a body with the sheet, is due to a natural chemical reaction, perhaps similar to the effect of herbaria leaves (Volckringer, 1991);

- The image was the result of the emanation of ammonia vapour (Vignon, 1902) or the interaction of gases produced by the corpse with substances derived from retting of the linen (Rogers, 2002; Rogers 2005).

- The image is a painting: many techniques have been proposed, but the best results were obtained using a modified carbon dust drawing technique (Craig & Bresee, 1994).

- It was obtained from a warmed bas-relief (Ashe, 1966; Pesce Delfino, 2000).
- It was obtained by rubbing a bas-relief with pigments or acids (Nickell, 1991).
- It was obtained by exposing linen in a darkened room using chemical agents available in the Middle Ages (Allen, 1993).
- It was a natural phenomenon a corona discharge – which occurred in the grave during an earthquake (Lattarulo, 1998; De Liso, 2002).

These hypotheses may be grouped into four categories:

1. *energy acting at a distance;* the hypothesis of a radiation source coming from inside the enveloped body is the most probable, although, due to the need to manage with relatively high sources of directional energy, no complete experiments have been done to show physical results. From a theoretical point of view all the particular characteristics of the TS could be reproduced if a corona discharge[1] is supposed (Fanti, Lattarulo & Scheuermann, 2005). Some problems still arise if other energy sources like proton are supposed because Items G9 and O3, O4 must be verified.
2. *body-sheet contact:* the body image is due to a chemical process similar to that which happens in leaves of herbaria (Volckringer, 1991); the proposed mechanisms involving direct body-sheet contact, are inconsistent with Items G9, G10, G15, G16 and O2, O3, O5.
3. *a diffusion mechanism:* caused by the ammoniacal gases of decomposition emanated from the corpse enveloped in the TS; this hypothesis, is inconsistent with Items G4, G9, G10, G12, G16 and O2, O3, O4, O5.
4. *intervention by an artist:* Some examples are the following.
   - The body image is a painting; perhaps the best results was obtained using a modified carbon dust drawing technique (Craig & Bresee, 1994);
   - it was obtained from a warmed bas-relief; tests involved scorching a sheet laid on a heated bas-relief (Pesce Delfino, 2000);

---

[1] The body image could be due to an energy source of protonic, UV or other types, but more probably the energy source was related to a corona discharge. The corona hypothesis explains many facts, some of the most interesting of which are: the need to assume the soft hair; that there was a radiation source normal to the skin; the absence of detectable defects on the cellulose crystals after the energy application; the uniform color of the thin layer of polysaccharides around the linen fiber; the discontinuous color along the yarn; the 3-D information; the double superficiality of the image.

- it was obtained by rubbing a bas-relief with pigments or acids (Nickell, 1987);
- it was obtained by exposing linen in a "darkened room" using chemical agents available in the Middle Ages (Allen, 1993).
- Apart from the conjecture that a primitive kind of photography existed in Medieval times, the general hypothesis of intervention by an artist who constructed the image artificially is inconsistent with Items G2, G4, G11, G13, G16, G19, G20 and O5.

It must be observed that some researchers (McCrone, 1997) have detected the presence of iron oxide particles and also found tempera on the image area of the TS, hypothesizing that these substances may have contributed to image formation, but the STURP (Jackson, 1984, Jumper at el., Schwalbe) presented several arguments opposing this hypothesis, because such traces are not sufficient to revive the color of the area in question. One argument is that no evidence of capillary flow between the cloth fibers has been reported, and X-ray fluorescence did not detect substantial differences in iron concentrations between image and non-image areas. Mass spectrometry and ultraviolet fluorescence failed to detect the presence of paint traces in significant image areas. Furthermore, various painters copied the TS through direct superimposition and some traces of paint may have remained.

Clearly, any scientist who makes a statement in reference to the Resurrection leaves the domain of science, at least partially, since a scientifically unexplainable phenomenon is presumed to exist. Although good experimental results have been obtained in the sense that, at first sight, the image, generally limited to the face, is similar to that of the TS Man until now no experimental tests have been able to reproduce all the qualities found in the image impressed on the cloth (Fanti, 24 authors, 2005).

From a scientific point of view, it is very interesting to understand how a corpse could have generated such a peculiar image even now not reproducible in all its characteristics. From a religious point of view, it is important to understand what the TS is, because, if it is authentic, it witnessed the event of the burial and Resurrection of Jesus Christ.

1.6. Optical Research: Limits and Perspectives

Such a particular body image imprinted on the TS, must be studied from a multidisciplinary point of view not excluding anthropology, biology, chemistry, computer science, engineering, geology, history, measurement science, medicine (in many different branches such as anatomy, pathology, and forensic medicine), microbiology, palynology, photography, physics, theology and many others disciplines, but perhaps the most important and promising is the optics aided by the computer science.

The in depth optical study of high resolution images, corroborated by the others disciplines, will probably be the best way to try to find an explanation for the most important question: how the body image formed on the TS? This will be achieved if high-resolution quantitative photographs will be analyzed by means of ad-hoc optical methods, but to reach this goal, a problem must be solved. The tests on the TS must be very limited because the body image is photosensitive and degradable: as the image is due to a precocious ageing of the linen fibers, any ambient factor such as temperature and light increases the ageing process of the TS background and therefore the contrast image/background tends to drastically annihilate, reducing the capability of detecting the presence of the image. In addition the Turin Archdiocese is very close to new test and it is not easy to obtain the permission to apply a particular test procedure.

In this view, it is not easy to control the measurement repeatability of the measurements done, necessary condition to have reliable data, because once a test is directly performed on the TS, it is not easy to have the permission to control data especially in the cases when some results can be questionable.

For example in 1978 some photometric measurements were directly made on the TS (Artom & Soardo, 1981), but once the acquired data were processed, some results were doubtful. For example the MLF (Mean Luminance Factor) of the background (0.50 with a standard deviation of 0.09) was too close to the body image MLF (0.50 with a standard deviation of 0.07); for tens of years it was then impossible to verify and improve these measurements because no further direct access was allowed.

For this reason the future research on the TS will be developed on the basis of photographic material collected in the past with the application of new optical methods developed for this purpose; in particular non standard optical procedures must be studied and applied to the relatively few photographs put at disposition of the scientific world. A particular task for optics and photography will be needed: that to design very reliable optical techniques that will allow to obtain a big quantity of information with minimal errors in very few time of acquisition in order to disturb as less as possible the body image of the TS.

After a review of the Optics research done in the past and in these years, some possible techniques and methods to be applied in the next future will be discussed both to try to understand how the body image formed and to monitor the conditions of the TS for conservation purposes.

2. OPTICAL RESEARCH: THE PAST

This section lists in eleven paragraphs the most important results obtained by the optical research in a century starting from the first photographs obtained in 1898 and arriving to the image processing performed before the third millennium.

2.1. The Negativity of the Body Image from the Photographs of S. Pia of 1989

It can be stated that the optical research on the TS began in 1898 with the first photograph made by Secondo Pia (Pia) (1855-1941) on May, 25, 1898. He made the first two photos experimenting with a new technique for that period. He used two electrical headlights covered by two polished glassed to better diffuse the irradiated light.

After the first two photos the heat emitted by the lights broke the glasses and the work stopped up to May, 28. The two b/w orthochromatic plates 50x70 cm large (produced by Edward) had an exposure respectively of 14 and 20 minutes using a Voigtländer objective with a two millimeters diaphragm and a light-yellow filter. When developed the plates showed the Face of the TS of figure 2.1 and the front-and-back body image of figure 2.2.

For the first time it was therefore possible to admire the face of the TS Man and the whole double image of a corpse with the tone not reversed as they are on the Relic: in other words the body image appeared better distinguishable to the brain-eyes human system because it was imprinted in the linen Cloth as if it were a negative image.

Even if the body image is not a true negative of the reproduced body, it globally appears as such, in a first approximation. For example the hair that in the Pia's plate are almost white, then typical of an old man, they were not probably of that color. This result was achieved on the Relic because all the anatomic elements enveloped in the Cloth interacted with the linen fabric in a similar way independently from their own color. The most probable effect that caused that image, as it was previously discussed, was the corona discharge (Fanti, Lattarulo & Scheuermann, 2005): in this context it must be evidenced that the point effect (that yields in correspondence of little curvature radii) which causes a local increase of the electrical discharges intensity, well explains why the hair, geometrically constituted by many cylinders having very little radii, imprinted their image better than many other body parts.

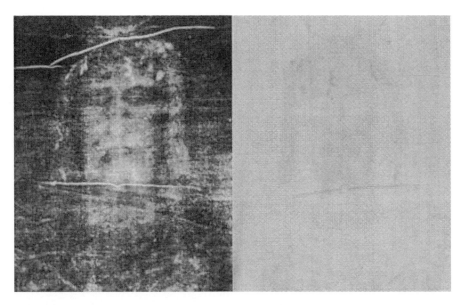

Figure 2.1. On the left first negative photograph of TS Face made by S. Pia in 1898 compared with a the face as it can be seen looking at the TS (B. Schwortz 1978).

Figure 2.2. First negative photograph of the whole TS made by S. Pia in 1898 (sizes of 53 x 14 cm).

On the other hand, the bloodstain that were directly impressed by contact on the fabric, appears as they were in positive on the TS and negative in the Pia's negative photographs.

2.2. The Orthochromatic Enrie's Photographs of 1931

The negativity of the body image discovered by Pia was hardly put in discussion by researchers that were against the authenticity of the TS: they accused Pia to have voluntary made a false photograph even if from an optical point of view it is not easy to think a way to make such an artifact. It must be waited for 33 years for the verification of the negativity when Giuseppe Enrie made a second series of photos in occasion of the big ostension of the TS of 3-24 may 1931.

During the night of May, 3, 1931, two hours were allowed to Enrie for the acquisition of 12 photographs (Enrie, 1933) in 30x40 cm and 40x50 cm plates of the whole TS, the Face, the back of the Body and the Wrists. The Relic was illuminated by means of lamps for a total of 16000 candelas, the photographic plates were of orthochromatic type filtered by yellow screens and the exposure time varied from 1.5 to 9 minutes.

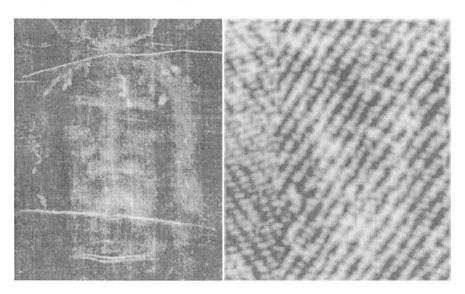

Figure 2.3. photograph of the TS face made by G. Enrie in 1933 and on the right a detail of the cloth herringbone pattern in correspondence of the nose showing the relatively high resolution obtained.

Enrie chose a b/w orthochromatic plate to acquire the photos because he wanted to enhance the relatively low contrast of the body image on the background; furthermore it is known that an orthochromatic plate is more sensitive to the red-brown than to the cold colors and the TS body image just has such colors. The orthochromatic photographs made by Enrie, see figure 2.3, reached a relatively high resolution because they were capable to resolve also the 3:1 herringbone weaving of the linen Cloth; as it is shown in figure 2.4 that represents the Fourier power spectrum of the Face, there are picks in the range from 0 to 100 m^{-1} typical of the body features and peaks at 630 and 980 m^{-1} that correspond to the texture. This high resolution of the photographs pushed many researchers in the very close analysis of the details of the TS that leaded to many interesting discovers, unfortunately not all completely reliable. For example the signs of the Roman flagrum on the tortured body were well evident, see figure 2.5 and it was possible to detect that more than one hundred of blows were inflicted to the victim by at least two different scourgers one posed at the right and one posed at the left of the Man.

Nevertheless someone, perhaps neglecting the aliasing problem due to the grain effect of the glass plate (the spatial resolution of the photograph is comparable with the texture dimensions), detected signs as microscopic letters attributed to names of hypothetical coins that probably are only alias due a sampling problem (see § 2.8 and 2.9).

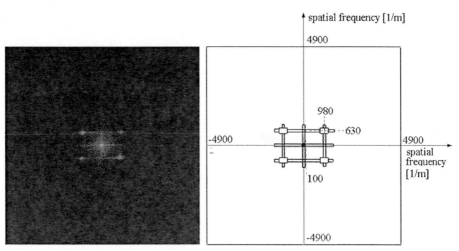

Figure 2.4. bidimensional Fourier Transform (power spectrum) of the photograph of the Face of figure 2.3 (on the left) and sketch of the same plot (on the right) that evidences clearer spatial frequencies at 980 and 630 m-1 that correspond to the herringbone texture (Fanti & Maggiolo, courtesy of JOPA).

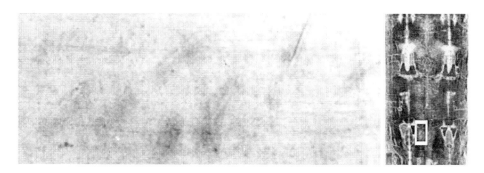

Figure 2.5. zoom at contrast enhanced of a photo of Enrie in which the signs of the scourges in correspondence of the left leg dorsal image are shown; on the right, corresponding position of the detail in the dorsal negative image.

2.3. The Multi-Spectral Judica Cordiglia's Photographs of 1969

Despite the big interest provoked by the optical research on the Enrie's photographs, other 38 years passed before to have the possibility to control the results obtained by G. Enrie. In 1969 Gian Battista Judica Cordiglia (Judica Cordiglia, 1974; Judica Cordiglia, 1988) had the possibility to make a set of photos during the private recognition of the TS on June, 16, 1969.

The objective of Judica Cordiglia was not that to reach a resolution better than that obtained by Enrie but he pointed to the acquisition in different spectral conditions; the following series[2] of photograph were made, by means of a Mamya C 220 camera having an aperture of 11 or 22 and exposures varying from 1/5 s to 10 s: negative colors; positive colors; negative b/w; negative reflected UV; negative UV filtered to acquire the only fluorescence; negative IR, see figure 2.6.

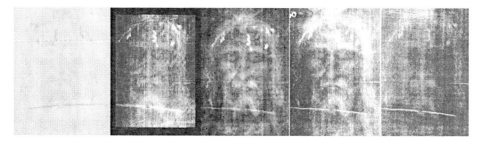

Figure 2.6. TS Face respectively made in positive colors, negative b/w, negative reflected UV, negative filtered UV, negative IR [courtesy of Judica Cordiglia].

---

[2] For series it is intended the photograph of the whole TS and some details of Face, Hands and Feet.

It is interesting the comparison of the photographs of figure 2.6 because some details evidenced in a particular spectral condition are not reported in others; for example the "3" bloodstain on the forehead and the moustaches are quite different if seen in IR light; the tumefaction on the right cheek is more evident in the UV photos that reveals more superficial details. Figure 2.7 that compares the hands acquired in reflected UV and in IR light, evidences the pronounced superficiality of the body image that is much better detected by electromagnetic waves having shorter wavelengths such as the UV. Perhaps a deep analysis of these characteristics will help in the understanding how the body image formed, but the non uniform conditions of illumination on the TS adds some difficulties.

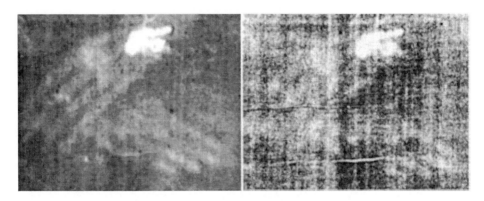

Figure 2.7. photograph of the TS Hands in reflected UV light, on the left, compared with the same image obtained in IR light [courtesy of Judica Cordiglia].

2.4. The M. Frei Microscopic Analysis of Sticky Tapes of 1973

It must not be forgotten that also microphotography gave useful information for the TS and it past history. In 1973 M. Frei (Frei, 1979) collected some fibers and dusts directly taken from the TS by means of the application of sticky tapes put on predefined areas of the TS; using an optical microscope he first detected that the material withdrawn from the TS surface was composed of many different particles of linen fibers, image fibers, scorched fibers, blood and pollen that he classified as coming from Palestine and Turkey; in addition he also found recent pollen coming from Europe. From this data he concluded that the TS must have been in Palestine and in Turkey in agreement with the tradition and the history. An example of the micro particles contained in the TS is shown in figure 2.8.

Figure 2.8. example of some vegetal particles typical of dusts aspired from the TS by G. Riggi di Numana during the 1988 recognition and conserver at Fondazione 3M of Milano Segrate, Italy.

2.5. The P. Gastineau Photo-Relief of 1974

Considering the particular characteristics of the TS image, some researcher detected an apparent relationship between the grey tones of the photograph and the depth of the details represented in it; with the advent of the computer machines it was easier to show this appearance. In 1974 the engineer Paul Gastineau (Gastineau, 1986; Guerreschi; Legrand, 1998) invented a particular laser-based system derived from the reading of banknotes, capable to transform the grey tonalities of the Enrie's photo of the TS Face into a mould that well represented the 3-D characteristics of the TS Face: that was the first demonstration that the Face has encoded some 3-D features that are not typical of other photographs of human faces, see Figure 2.9.

Figure 2.9: first 3-D mould of the TS Face made by P. Gastineau in 1974 (courtesy of Jean Dieuzaide in [Gastineau]).

The relationship between the image intensity and the cloth-body distance was successively verified by J. German (German, 1977): who determined the equation that best describes the image intensity, in terms of relative opacity of the photograph R_O, versus cloth-body distance D as:

$$R_O = 9 + 46e^{-1.03D} \qquad\qquad (2.1)$$

2.6. Some Optical Analyses Made by the STURP in 1978

In October 1978 the STURP (Shroud of TUrin Research project), a group of 32 American scientists (Jumper & Mottern, 1980) scientifically leaded by J. Jackson and organizationally by P. Rinaldi, operating with their own time and with their own funds, made the most important and reliable analysis on the TS ever done. The primary goal of their investigation was to determine the physical and chemical characteristics of the body image and they made a great number of non destructive tests that furnished many information very useful for the understanding of the peculiar characteristics of this particular Object.

After a very deep multidisciplinary analysis the STURP concluded that it was simpler to state what the TS body image is not than to state what it is. In fact they concluded that the image was not a painting but they were not able to explain how it could have been formed even if some tentative hypotheses were formulated: in synthesis these scientists agreed that what can be explainable from a chemical point of view it was not from a physical one and vice versa (Schwalbe & Rogers, 1982).

Here are reported and discussed some of the most interesting test done by STURP from an optical point of view. Tests include:

a. x-ray fluorescence spectrometry (Morris et al., 1980);
b. low-energy radiography (Mottern, et al., 1979);
c. Photoelectric spectrometry (Accetta & Baumgart, 1980; Gilbert & Gilbert, 1980; Pellicori, 1980);
d. Thermographic investigation (Accetta & Baumgart, 1980);
e. Reflected UV and Visible photography (Miller & Pelliconi, 1981; Miller, 1982);
f. Photography in transmitted and reflected visible light (Jackson et al.,1984);
g. Direct macroscopic visual inspection (Schwalbe & Rogers, 1982);

h. Direct microscopic visual inspection (Schwalbe & Rogers, 1982; Pellicori & Evans, 1981);

i. Microscopic analysis of sticky tapes (Schwalbe & Rogers, 1982);

j. 3-D features of the body image (Jackson et al., 1984).

2.6.a. X-ray fluorescence spectrometry

The most important non destructive test to detect the possible presence of inorganic pigments was the x-ray fluorescence (Morris et al., 1980). A painting produced from an iron, arsenic, lead or other heavy metal compounds would have revealed its presence when comparing the body image with the background. In reference to elements having atomic number greater than 16, the system employed demonstrated that there were no detectable differences between image and non image areas; it can be therefore concluded that the image was not painted with inorganic pigments. Uniform distribution of calcium, strontium and iron were also detected and this presence may be correlated with the retting process of the linen fibers.

2.6.b. Low-energy radiography

The entire TS was radiographed (Mottern et al., 1979) in order to detect the possible presence of heavy-metal pigments, but no density discontinuities between the body image and the background were observed, confirming the results detected in § 2.6.a. Figure 2.10 shows an example of the images obtained during the test.

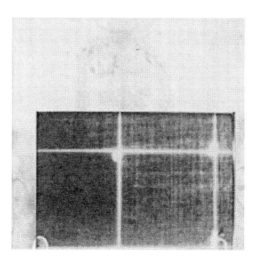

Figure 2.10. low-energy radiography of the Face area [Mottern] superimposed on the TS body image; no signs of body image result on the radiography.

2.6.c.   Photoelectric spectrometry

Accetta and Baumgart [1980] used a black-body emitter at about 1250 K and a HgCdTe detector to scan spots of 2 cm in the 3-5 μm and 8-14 μm band (figure 2.11) but their results were judged inconclusive because of the relatively large fluctuations in atmospheric absorption and the low reflectance to the cloth (5-10%).

Gilbert and Gilbert (Gilbert & Gilbert, 1980) measured UV and visible reflectance and fluorescence spectra from body image, scorch, bloodstain and background of the TS. The instruments allowed the data acquisition in the range from 250 to 750 nm with a resolution of 5 nm. For the reflectance measurement, a 150 W xenon lamp irradiated an area of 3 x 6 mm. For the fluorescence measurements a 200 W mercury arc lamp with the source monochromator set at 365 nm was used; they also placed an UV-transmitting/visible-absorbing filter in the source beam to eliminate the visible radiation incident to the target and they obtained data continuously with a bandwidth of 8 nm.

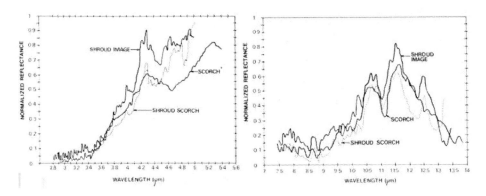

Figure 2.11. normalized spectral reflectance comparison of scorched linen with averaged Shroud image and scorch areas in 3-5 □m band (on the left) and 8-14 □m band (on the right) [Accetta, courtesy of AO].

They showed that the background cloth areas fluoresce in a broad band with a maximum at 435 nm, when pure cellulose fluoresce very weakly. The body image itself did not fluoresce measurably and the image reduced the background fluorescence also shifting the maximum to longer wavelengths.

The Gilberts detected that the relative reflectance of the body image increases with wavelength increasing and it resulted that none of the spectral characteristics expected from normal dyes, stains or pigment can be observed in correspondence of the TS image. Figure 2.12 and figure 2.13 respectively show the relative spectral reflectance and fluorescence of TS body image areas.

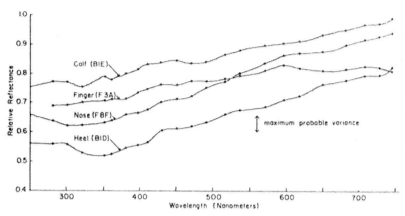

Figure 2.12. UV-visible relative spectral reflectance of four TS body image areas: calf, finger, nose and heel [Gilbert, courtesy of AO].

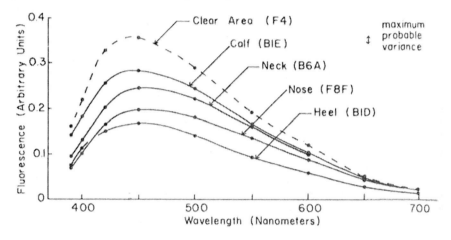

Figure 2.13. spectral fluorescence with excitation at 365 nm of four TS body image areas (calf, finger, nose and heel) compared with the TS background [Gilbert, courtesy of AO].

Pellicori (Pellicori, 1980) made similar measurements using a 500 W tungsten that illuminated a sample of 1 cm^2; he reported an approximate ±1% repeatability for the measurements and a spectral resolution of 17 nm reading out digitally at 20 nm intervals in the range of 440-700 nm.

The two different results obtained by Pellicori and the Gilberts were compatible within a range of 5%.

2.6.d. Thermographic investigation

Accetta and Baumgart (Accetta & Baumgart, 1980) observed possible inhomogeneities in the TS image attributable in IR emissivities not otherwise detectable in the visible region. Imaging was accomplished in the 3-5 μm and 8-14 μm bands with thermographic scanner cameras; the source was two 1500 W photographic floodlamps. With this illumination considerable contrast was noted in the 8-14 μm band but no feature were observed in the 3-5 μm band.

They concluded that the image observed in the IR and in the visible region was approximatively the same, but with reversed contrast: hands, face and feet appear relatively bright in the IR region as opposed to the visible appearance; as an example figure 2.14 shows the Face and figure 2.15 shows the hands, both acquired in the 8-14 μm band.

An aspect not yet observed that seems very interesting is the lack of image in correspondence of mouth and cheeks; this one should be certainly taken into account for the body image formation hypothesis that must also be able to explain this fact; for example it is be quite strange that an hypothesis based on the gas diffusion (Rogers, 2002) should be able to explain the lack of image just where it should be better impressed; an hypothesis based on corona discharge (Fanti, Lattarulo & Scheuermann, 2005) instead could explain this fact in reference to the void relative to the oral cavity.

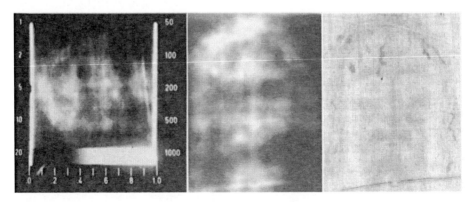

Figure 2.14. on the left, Face acquired in the 8-14 □m band [Accetta, courtesy of AO]; on the center smoothed and contrast improved photo of the same Face to be compared with the contrast enhanced photo of the Face made by B. Schwortz in visible light. In addition to the reversed contrast detected by the authors in the IR photo, it must be also observed the lack of image near mouth and cheeks.

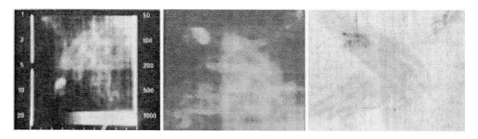

Figure 2.15. on the left, Hands acquired in the 8-14 ☐m band [Accetta, courtesy of AO]; on the center smoothed and contrast improved photo of the same Hands to be compared with the contrast enhanced photo of the Hands made by B. Schwortz in visible light. In addition to the reversed contrast detected by the authors in the IR photo, it must be also observed the different shape of bloodstains perhaps indicating the various degree of blood imbibitions and the presence of some stains on the palm that must be explained.

2.6.e. Reflected UV and Visible photography

Miller (Miller & Pellicori, 1981; Miller 1982; Jumper & Mottern, 1980) made mosaics at 5.6:1 and 22:1 reduction of the entire surface of the TS. For each section a successive series of exposures were made with red, green, and blue filters for color separation. In another series UV filters were used for contrast enhancements. To detect fluorescence in a different series, UV transmission filters were used over the light sources while UV blocking filters were used over the camera lens. For another series the visible spectrum was partitioned into 10 nm intervals by a series of filters (see figure 2.16).

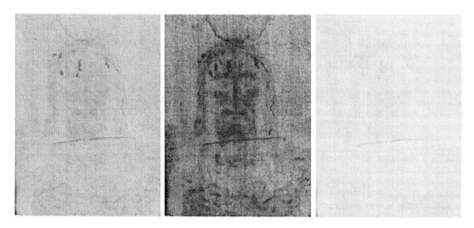

Figure 2.16. TS Face respectively filtered from left to right with green, yellow and red filters.

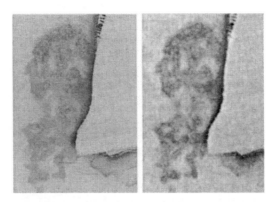

Figure 2-17. Chest wound and a TS patch photographed in visible light (courtesy of G.B. Judica Cordiglia on the left) compared with the one made in UV light (courtesy of V.Miller on the right) where the reddish fluorescence of the scorches is evidenced.

It was confirmed that the body image has no color if seen in UV light when the scorches showed a brownish-red fluorescence (figure 2.17) and the fluorescence of the serum areas was highlighted by circles of yellow-green associated to some wounds on the right side of the TS as shown in figure 2.18.

The multispectral narrow band Miller's photographs are qualitatively consistent with the spectra acquired by Pellicori and the Gilberts and show a decreasing image contrast from the UV to the red-visible.

Figure 2.18. the yellowish fluorescence of the serum halo is evidenced by the arrow in correspondence of the wrist wound (contrast enhanced photo of V. Miller)

2.5.f.   Photography in transmitted and reflected visible light

Barrie Schwortz documented photographically the whole activity done by the STURP, but his major contribution to the knowledge of the body image impressed on the TS consisted in the color photographs of the whole TS made both in reflected and transmitted light. Figure 2.19, acquired in reflected light must be compared with figure 2.20 acquired in transmitted light: the well evident body image even if not so much contrasted in figure 2.19 almost disappears in figure 2.20 showing that the image is very superficial (Jackson et al., 1984); details of contrast enhanced Hands shown in figure 2.21 confirm this fact.

Figure 2.19. whole TS in reflected light mounted on a rotating table used by STURP members for experiments (B. Schwortz).

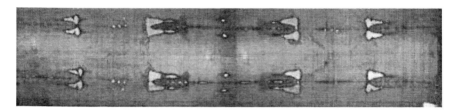

Figure 2.20. whole TS in transmitted light: the very superficial body image disappears but not the bloodstains and the waterstains; behind the TS it was sewed the Holland cloth that acted as a light diffuser (B. Schwortz).

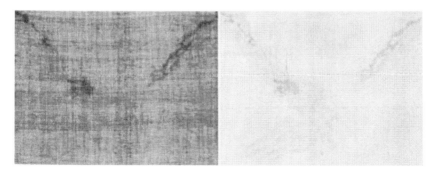

Figure 2.21. contrast enhanced photograph of TS Hands in transmitted light (on the left) compared with the same photograph in reflected light: the very superficial body image disappears but not the bloodstains on the wrist and arms
(B. Schwortz, image processing by G. Fanti).

2.6.g.   Direct macroscopic visual inspection
The TS was scrutinized in situ by many scientist and some conclusions are here reported (Scwalbe & Rogers, 1982): the water used to extinguish fire in 1532 migrated to the cloth in both scorched and unscorched image areas; the TS was subjected to very steep thermal gradients during that fire; there is no appearance of brush marks; the TS is characterized by bands of different color intensity probably also due to the use of a different material to weave the TS (linen coming from different retting processes); the image density changes at a location where it is evident the banding effect see for example figure 2.21.

2.6.h.   Direct microscopic visual inspection
The TS was directly analyzed with a microscope by Evans (Pellicori & Evans, 1981) and studied by Rogers (Schwalbe & Rogers, 1982) who used a 10x lens and a needle. Evans made a series of interesting photomicrographs that showed the details of the image-yarns: in particular it was shown that the color of the image-areas has a discontinuous distribution along the yarn of the cloth: striations are evident (Fanti, 24 authors, 2005). The image in fact has a distinct preference for running along the individual fibers making up a yarn, coloring some but not others, see figure 2.22.

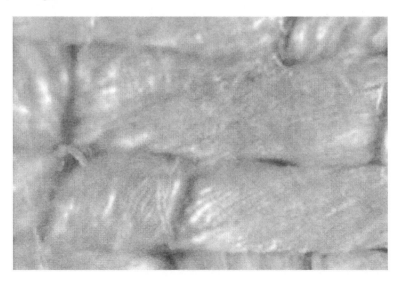

Figure 2.22. STURP-ME-29, photomicrograph of the image area in correspondence of the nose; the discontinuous distribution of the color is evidenced by the striations (courtesy M. Evans).

R. Rogers analyzed the nose area and detected that the image only resides on the topmost fibers of an image linen yarn; he detected that at maximum the color interested the upper three fibers in the yarn thickness (a TS yarn is composed of 80-120 fibers). This fact, confirmed the superficiality of the body image detected by B. Schwortz in §2.6.f.

2.6.i. Microscopic analysis of sticky tapes

R. Rogers and R. Dinegar applied to the TS (Schwalbe & Rogers, 1982) and the adjoining Holland cloth 32 sticky tapes, each about 5 cm^2 in area, with a roller specially designed for this purpose. These tapes were put in a special box for further microscopic analysis.

They partially confirmed the results of M. Frei of § 2.4, but they found a minor quantity of pollen perhaps due to the different sampling method. They used the red particles contained in these tapes to demonstrate that the red stains are of human blood (Heller & Adler, 1980) and the image fiber to make chemical analyses on them.

For example it was detected that the cellulose of the medullas of the fibers in image areas is colorless because the colored layer on image fibers can be stripped off, leaving colorless linen fibers (Fanti, 24 authors, 2005) see figure 2.23, and phase-contrast photomicrographs evidenced the characteristic typical of very old linen fibers as shown in figure 2.24.

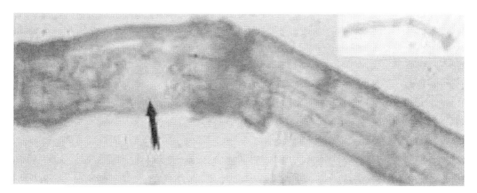

Figure 2.23. mechanically treated image fiber (sample STURP-1EB coming from the leg area, about 12 μm in diameter) showing a colorless medulla in correspondence of the arrow where a part of the external layer is absent. On the right it is also visible a small blood crust adhering the fiber (G. Fanti).

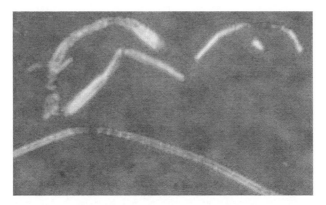

Figure 2.24. image fibers (sample STURP-1EB) in cross-polarized light showing the presence of defects and internal stresses along their length (G. Fanti).

It also resulted that there is a very thin coating on the outside of all superficial linen fibers on Shroud samples named "Ghost" (Fanti, 24 authors, 2005); they are colored (carbohydrate) impurity layers pulled from a linen fiber by the adhesive of the sampling tape and they were found on background, light-scorch and image sticky tapes, see figure 2.25.

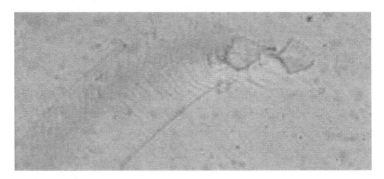

Figure 2.25. colored impurity layer, named "Ghost" [Fanti, 24 authors], pulled by the sticky tape from a linen fiber (sample STURP-1EB) in cross-polarized light (G. Fanti).

2.6.j.   3-D features of the body image

J. Jackson [1984] considered the body image with respect to the type of its spatially distributed shading structure. From the spectral data of Gilbert and Gilbert (Gilbert & Gilbert, 1980) and Pellicori (Pellicori, 1980) it was estimated that the absolute reflectivity $r(x, y, \lambda)$ of the TS body image at a point (x, y) and a wavelength l (from 300 nm to 700 nm) can be empirically characterized as:

$$r(x, y, \lambda) = (-0.21 + 0.00102\ \lambda)\ [1+0.00078\ (\lambda\text{-}790)\ h(x, y)\ ] \qquad (2.2)$$

being $0 \le h(x, y) \le 1$ (0= cloth background, 1= tip of the nose) the spatial variation of reflectivity that gives the appearance of a human form within an accuracy of 4%.

At an arbitrary wavelength the reflectivity is a linear function of $h(x, y)$ so that the reflectivity monotonically decreases (image become darker) as h increases. Thus the spatial appearance of the TS image is similar at all wavelengths although the dynamic range of reflectivity variation is greater at shorter wavelengths: as a consequence many studies of shading structure can be made from a single b/w photograph of the TS. Relating the reflectivity to exposure $E(x,y)$:

$$E(x,y) = \int_0^\infty s(\lambda)\, r(x, y, \lambda)\, L(\lambda) \cdot d\lambda \qquad (2.3)$$

where $s(\lambda)$ is the spectral sensitivity and $L(\lambda)$ is the illumination spectrum and relating the photographic density d to exposure (d versus $logE$ curves) and transmittance I:

$$d = K\ \log_{10} \frac{1}{I} \qquad (2.4)$$

J. Jackson [1984] showed that the transmittance, intensity and shading are a direct function of the spatial variation of reflectivity h.

From Eq. 2.2 it was calculated that at 550 nm the minimum absolute reflectivity (at the nose with $h=1$) is 28.5% and TS background 35.1% for a base ten logarithmic density range of 0.09 (a new linen has an reflectivity of about 58%). Image reflectivities at 680 nm (red) and 440 nm (blue) at maximum shading ($h=1$) are in the ratio 2.5/1 and for the background ($h=0$) 2.0/1, thus the body image appears on a yellowed linen as a faint but darker yellow-brown discoloration being h the only characteristic which varies from point to point as to produce the pattern of a human body.

The discoloration of the body image shading I was associated with a property $P(x', y')$ of a human body as:

$$\text{I}(x, y) = \text{f}\ [\text{P}(x', y')] \qquad (2.5)$$

being (x, y) the coordinates of some image point and (x', y') the coordinates of the associated image on a body shape. The property P was chosen as distance D between the TS and a human body.

The authors of the paper (Jackson et al., 1984) measured the transmittance of a b/w transparency of the face taken from a photograph made by Miller in 1981 using a microdensitometer and the body-cloth distance D in 13 data points by draping a linen model of the TS over a bearded volunteer lying prone on the x-y plane: they concluded that "some linear correlation [of 0.60%] with the image shading and cloth-body distance is present in the Shroud".

Figure 2.26. Cardboard model of derived VP-8 frontal image of the TS (Jackson et al.,1984, courtesy of AO).

The authors (Jackson et al., 1984) also scanned TS images with a VP-8 Image Analyzer, a system that displays images shading (transmittance of a given transparency) as a spatial relief; even if this procedure contains some elements of subjectivity, this technique is a reasonably good indicator of distance correlation with transmittance. Figure 2.26 shows the cardboard model of the relief surface of TS frontal body image processed by VP-8: this seems to closely correspond to a body shape. Also the Face shown in figure 2.27 demonstrates the "apparently high correlation of image intensity with the distance". The VP-8 image well modelled relief variations of a human form over a small scale (about 10 cm), but not over a large scale because the image resulted flat; in addition, it was also noted in plane distortions consistent with cloth drape (Ercoline et al., 1982).

Figure 2.27. VP-8 relief result (on the right) obtained from the TS Face (on the left) TS (Jackson et al., 1984, courtesy of AO).

These results demonstrate that the TS body image has 3-D characteristics that correlates two surfaces: the body shape and the cloth.

2.7. Photometric and Colorimetric Measurements by Artom & Soardo in 1978

Artom and Soardo (Artom and Soardo, 1981), in 1978 first measured both the three-chromatic coordinates (x, y, z) and the luminance factor of the TS in 71 spots having a diameter of 13 mm, employing a colorimeter Spectra-Pritchard-Photometer and a halogen lamp having a color temperature of about 3100 K coupled with two gelatine filters Kodak CP-05-Y to obtain a color temperature of 2856 K.

For example the averaged x, y, z values of three-chromatic coordinates in correspondence of the background respectively resulted 0.474 (0.004), 0.422 (0.001), 0.104 (0.005) (being in the brackets the standard deviation) to be compared with those of the body image that resulted respectively of 0.480 (0.006), 0420 (0.003), 0.100 (0.004). The measured luminance factor of the background was 0.50 (0.09) and that of the image 0.50 (0.07).

It is quite strange that all the three-chromatic values of the background and of the body image are compatible within the standard deviation and that the corresponding luminance factors are equal, but obviously there results were obtained after the test on the TS and the authors were not allowed to repeat the test immediately after to control their result; these data show the need of a proper calibration of the optical instrumentation before the use during the test that needs a particular attention in the case of the TS when rarely the experiment can be repeated. For this reason more robust methods and instruments than in other cases must be used when data are acquired from the TS.

2.8.   Presence of Coins Detected from 1982?

With the diffusion of computers it was very easy to study the fine details of a digitized image; many scholars analyzed the details of the photographs starting from those of Enrie that, being high-contrasted, seem to give the greater number of information.

A 3-D analysis shows a protuberance in correspondence of the right eye of the TS Man that is coherent with a corona discharge image formation (Fanti, Lattarulo & Scheuermann, 2005), but this protuberance can be also interpreted as the presence of a rounded object over the eye; F. Filas (Filas, 1982) first correlated the possible presence of a coin with the imprint of the Dilepton Lituus coined under Pontius Pilate in 29-30 A.D. (from the letters "IS": I=10, S=6; sixteenth year of the Emperor Tiberius) see figure 2.28. This, because he found the shape of the "lituus" very similar to that detectable on the TS photograph of Enrie. Further studies verified the presence of this coin because, after having distinguished four letters "UCAI", Filas (Filas, 1982) and M. Moroni (Moroni, 1988) identified these as part of the inscription "TIBERIOU CAICAPOC" with the error CAICAPOC instead of KAICAPOC reported in some coins.

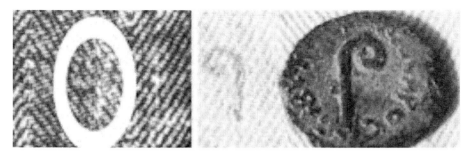

Figure 2.28. Imprint of the lituus on the TS photograph of right eye (on the left) compared with the imprint of a Dilepton Lituus on a TS-like sheet (on the right) (Courtesy of M. Moroni)./

These interesting result is nevertheless put in discussion by the possible aliasing effect (impossible to be eliminated a posteriori) due to the grain of the Enrie orthochromatic photographs that highly enhanced the contrast; the grain dimensions are in fact comparable with those of the TS yarns, those of the supposed "UCAI" letters and of the lituus. To show this effect, figure 2.29 compares a detail of Enrie orthochromatic plate with a macrophotograph, made by Mark Evans during STURP campaign in 1978, and an image processing of it devoted to enhance the contrast in such a way as an orthochromatic plate normally does. This figure shows how the herringbone structure of the TS changes in the Enrie's photograph as a combination of black sticks and how much a combination of the resulting black sticks can be interpreted as the letters I, C, U and also A. On the other hand, if someone searches on the whole TS photograph for those letters he is able to find more than some hundreds.

Figure 2.29. detail of Enrie's orthocromatic plate (on the left) compared with a similar detail of a macrophotograph taken by M. Evans (in the center, STURP-ME-18) that shows the herringbone 3/1 warp; the image processing of the same detail of M. Evans (on the right) after blurring and contrast enhancement appears similar to the detail of Enrie's plate. This processing shows how the details were lost in Enrie's photographs.

The repeatability of the acquisition was verified considering the results of two different photographs made by Enrie, but more recent, higher resolution photographs of the TS do not show the clear presence of these letters: the reproducibility, in reference to the results of other acquisition systems, is therefore not yet verified. For the moment the possible presence of coins on the TS Face must be then considered questionable.

2.9. Presence of Inscriptions?

A. Marion (Marion, 1998) improved the findings of many other scholars who detected the presence of many inscriptions around the TS Face such as "Nazarenus", "IC", "In Nece", see figure 2.30. To detect such letters Marion scanned the high resolution Enrie's photograph of Face and other made by V. Miller.

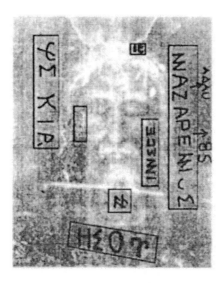

Figure 2.30. inscriptions detected by A Marion in correspondence of the TS Face [courtesy of Marion].

To maximize the signal to noise ratio he eliminated the herringbone double pattern present in the photographs "without altering the information" and he concentrated in a single output image all the pieces of information derived from several pictures of the same area corresponding to different conditions of acquisition and of digitization. After this he made some image processing such as dilatation, erosion, diffusion and smoothing by means of proper convolution filters, frequency filtering in the Fourier domain to evidences the possible presence of letters in the face photograph.

As it was discussed in the previous § 2.8, also in this case the real presence of letter is questionable for the following reasons: also some aliasing problem connected with the spatial sampling frequency can be evidenced; a multiple image processing that uses dilation and erosion in series can generate alias images in a photograph that contains many kinds of disturbs (such as herringbone patterns, bloodstains, waterstains, photographic defects and so on); also in this case in the more recent, higher resolution photographs of the TS it is not clear the presence of these letters.

2.10. The Photo-Relief

Another technique that subjectively confirmed the 3-D characteristics of the TS body image is the photo-relief, a technique used by photographers to give the sense of three-dimensionality (Guerreschi); it consist in the overlapping (both physically of digitally) a positive transparency of a subject with the corresponding negative, properly contrasted and slightly displaced with respect to the positive one. In such a way the resulting shading in some cases furnishes an appreciable 3-D result.

A. Guerreschi in 2000 applied this technique to the photo of the TS making a comparison with a normal photo of face: a person was photographed near photo of the TS and using the photo-relief technique, a subsequent processing was made; the same photograph was also processed with the VP-8 system described in § 2.6.j and the results are reported in figure 2.31. While the TS Face shows some 3-D characteristics in both cases, a normal photo of face has not this peculiarity.

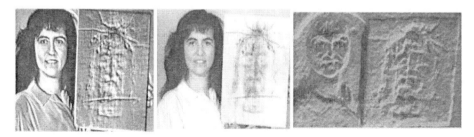

Figure 2.31. In the center a photograph of a person near the TS Face; the photo-relief technique (on the left) evidences the 3-D characteristics of the TS Face but not those of a normal face; the same conclusion can be achieved if the same photo is processed by means of VP-8 (on the right) [courtesy of A. Guerreschi].

2.11. The Polarized Image Overlay Technique

In 1985 (Whanger & Whanger) presented a new technique useful to compare the characteristics of different images: the two images in comparison are projected onto a suitable screen through polarizing filters and these superimposed images are then viewed through a third polarizing filter. The two filters connected to the projectors have their polarization axes at right angles, the two images are precisely superimposed on the screen by the aid of four references and the third polarizing filter is moved by the observer of a 90° arc to alternatively evidence one image in comparison with the other one; a time-varying comparison of both fine details and larger features is therefore possible because at extinction, the light transmission of two polarizing filters falls from about 22% to about 1%.

This procedure was successfully applied (Whanger & Whanger) to the comparison of the TS Face with the first coin bearing the facial image of Jesus Christ that measures only 9 mm from the top of the head to the point of the beard: a Byzantine coin of Justinian II (a gold solidus struck between A.D. 692 and 695), see figure 2.32.

This result is interesting because the authors found 145 congruence points between the two images analyzing position of bloodstains, configuration of the hair, nostrils, tonally shaded areas and replication of wrinkles on the TS; they therefore considered the image of Christ on the coin "identical to the Shroud". This fact makes therefore to suppose that the TS was visible in the Byzantine Emperor in the VII century A.D. about 600 years before the debatable radiocarbon TS dating of 1988.

At the end of the XX century this technique based on the cross-polarization of projected images was frequently substituted by the simpler technique based on the superimposition of digital images having a predefined degree of transparency; many software are able to vary with time the degree of transparency of two or more images superimposed each other (figure 2.32 shows an example).

3. OPTICAL RESEARCH: THE PRESENT

The present is here intended for studies published after 2000. Unfortunately after the debatable result of the TS radiocarbon dating obtained in 1988, many researchers devoted their interest in more urgent problems, letting the TS studies to future times in which the authenticity problem will be clearer. Also the Turin Archdiocese that is the responsible for the TS studies showed many closure signs and the direct research on this Relic was and it is more difficult.

In any case the research on the TS was not stopped for this result and this is demonstrated by the direct studies done in 2002; unfortunately most of these results were not yet made at the disposability of the scientific community. If the Turin Archdiocese is now very close, other researches have been very open and they have allowed some studies to go more in depth: for example Ray Rogers that during the STURP campaign of 1988 sampled many fibers and particles from the front side of the TS, made these samples free to the scientific community; Giovanni Riggi di Numana who sampled many dusts from the back side of the TS in 1978 and in 1988, through the Fondazione 3M of Milano Segrate (Italy) grouped these dusts with other TS samples and made them free to the researchers who asked for them; the first author of this paper takes the occasion to thank both R. Rogers and G. Riggi di Numana for the important samples freely sent to him.

Notwithstanding this closure also from an optical point of view some researches were done and some are still in progress; they are here reported.

Figure 2.32. on the center comparison of the facial image of Jesus Christ on a Byzantine coin (on the left) with the TS Face (on the right); the polarized image overlay technique used by (Whanger & Whanger) can be easily substituted by a digital superimposition of the images (transparency of 50% in the present case) as it was here done by G. Fanti using Paint Shop Pro 7.1® software.

3.1.   FTIR Microspectrophotometric Analysis

Starting from the end of the XX century, Alan Adler (Adler, 2002) took advantage of the new microspectrophotometric FTIR technique based on the Fourier transform of the optical signal acquired in the IR by an optical microscope to better characterize both the TS samples taken by R. Rogers during STURP campaign and the samples taken from the neighbourhood of the ^{14}C sample of 1988.

The FTIR analysis identifies chemical compounds in consumer products, paints, polymers, coatings, pharmaceuticals, foods and many other products. The resulting Fourier spectra in the wavelength domain, produce a profile of the sample, a distinctive molecular fingerprint that can be used to easily screen and scan samples for many different components; functional groups and covalent bonding can be clearly detected.

Adler verified that the FTIR spectra relative to fibers picked up near the ^{14}C area of the TS were not congruent with those of the TS picked up by R. Rogers during STURP campaign and first showed with experimental data that the 1988 sample was not representative of the whole Cloth; he also hypothesized a medieval patch invisibly sewed the TS, see figure 3.1.

He also verified that the FTIR spectra relative to the blood and serum particles were typical of human blood and serum spectra and that they differ very much from blood simulacra such as a dried blood and a mineral simulation see figure 3.1 confirming once again (Heller & Adler, 1980) that the red stains detectable on the TS are of human blood.

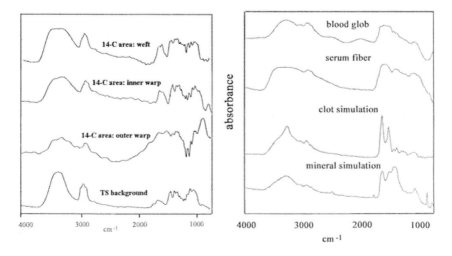

Figure 3.1. typical FTIR absorbance patterns of the TS background fibers compared with those relative to the 14C area (on the left) and typical FTIR absorbance patterns of TS blood and serum compared with blood simulacra (on the right).

3.2. Computerized Body Image Cleaning and Reconstruction

With the development of software relative to the optical processing, it was possible to improve the quality of the body image of the TS; to do so, the image was contrast enhanced, cleaned and some lacking parts were reconstructed.

3.2.a. Contrast enhancement

An RGB photograph of the TS presents the following luminance levels L that can be evaluated by means of the relation:

$$L = \sqrt{\frac{R^2 + G^2 + B^2}{3}}$$
$$(3.1)$$

being respectively R, G and B the luminance levels (in the range from 0 to 255) in the red, green and blue band.

As it can be seem in Table 3.1, the body image has a low contrast and to evidence some details of the body image, a contrast enhancement is required: in terms of luminance, the body image range is only 33 with a mean of 115 when around image the mean background is 164 (see Table 3.1). This contrast enhancement obviously also enhances the noise that is not negligible in the TS photographs.

For example, if we look at the interesting photomicrograph of nose of figure 3.2 we see that in absence of contrast enhancement (of 70 % coupled with a darkening of 33 points over 255) the important "striations" of the image fibers[3] are not so easy to detect.

Table 3.1. RGB and L values of some details of the 2000 TS photo [Archdiocese of Turin]

Image type	R	G	B	L
Background (min)	88	63	63	72
Background (near body image)	185	161	144	164
Background (max)	191	166	146	168
Background range	**103**	**103**	**83**	**96**
Body image (min)	138	112	91	115
Body image (max)	168	144	131	148
Body image range	**30**	**32**	**40**	**33**
Scorches (min)	56	38	37	44
Scorches (max)	170	141	119	144
Scorches range	**114**	**103**	**82**	**100**
Bloodstains (min)	127	96	83	103
Bloodstains (max)	158	132	111	135

---

[3] The striations are typical of the TS image and difficult to reproduce without a system of discharges caused by an electrostatic field (Fanti, Lattarulo & Scheuermann, 2005)

Bloodstains range	31	36	28	*32*

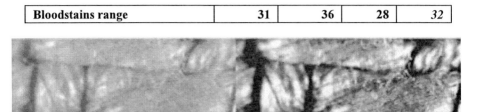

Figure 3.2. on the left photomicrograph of the image area in correspondence of the tip of the nose (courtesy of Mark Evans, STURP/ME-29); on the right contrast enhanced image.

3.2.b. Image cleaning

After the contrast enhancement, it was necessary to clean the amplified optical noise4 (Fanti & Faraon, 2000) of the body using eidomatic techniques capable to eliminate the bloodstains, the waterstains and other signs.

The Enrie's photograph of the TS, digitalized at high resolution (1440x5970 pixels) was cleaned by means of Paint Shop Pro 7.04® software; basing on the hypothesis that the TS background is uniform, the illumination non-uniformities were successively corrected varying the luminance in some selected areas of the body image.

The cleaned image of the TS was also used to evaluate the anthropometrical indexes and among them it was determined the tibio-femoral index of 83.8% that is very significant for the human body[5].

3.2.c. Image reconstruction

A kinematic analysis also based on the anthropometric indexes to characterize the dimensions of the enveloped man, allowed to determine the positions of the lacking parts such as arms and shoulders; the shoulder width defined by the points

---

[4] The optical noise consists in disturbs relative to the acquired image and some other marks present on the TS that they are interesting from other points of view but not for body image analysis purposes.

[5] It was also observed the compatibility between the indexes relative to the frontal and dorsal body image and they resulted very similar to that of the Semites (83,7 %) (Fanti, Marinelli & Cagnazzo, 1999); incidentally the tibio-femoral indexes relative to declared copies of the TS were also measured and they resulted greater than 100% showing the incompatibility with the human body characteristics.

A and *A** in figure 3.3 named akromion, was evaluated as 51.±0.5 cm The *q* and *q'* angles were determined solving the system of kinematic equations:

$$\begin{cases} x_A = BC \cos q + AB \cos q' \\ y_A = BC \sin q + AB \sin q' \end{cases} \qquad (3.2)$$

$$(3.2)$$

where the segments *AB* and *BC* were derived from the manikin characterized by the anthropometrical indexes previously evaluated. The position of point A was finally determined with a maximum uncertainty of 4 cm leading to the possible profiles shown in Figure 3.3 (on the right). The ϕ shoulder angles resulted: $\varphi_1 = 10 \pm 1°$ (right); $\varphi_2 = 12 \pm 1°$ (left).

The final body image (Cunico, 1999-2000) (see figure 3.4) obtained after reconstructing the lacking parts, cleaning and filling with colors corresponding to those of the body image, was analyzed to detect the characteristics of the TS Man there enveloped.

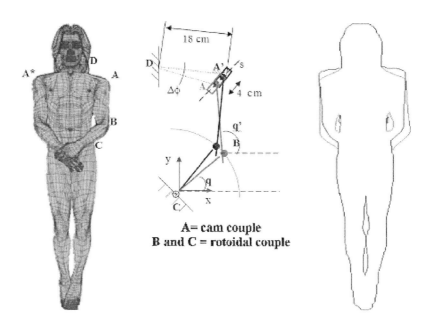

A= cam couple
B and C = rotoidal couple

Figure 3.3. on the left numerical manikin built in agreement with the detected anthropometric indexes; in the centre scheme of the kinematic analysis used to determine the profile limits of the TS Man (on the right).

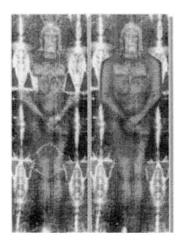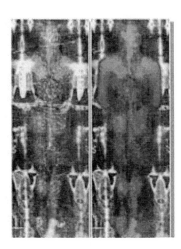

Figure 3.4. frontal (on the left) and dorsal (on the right) TS body image respectively as it is directly visible and after cleaning and reconstruction.

For example, the 3-D characteristics of the dorsal body image, that due to the presence of the scourge marks were not evident before this image processing, were confirmed in the processed body image (see figure 3.5).

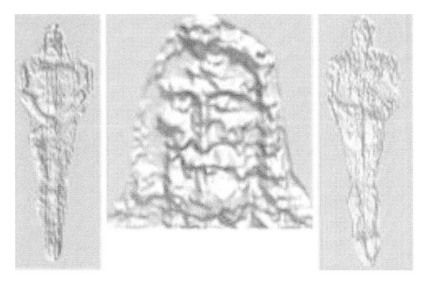

Figure 3.5. 3-D representation of the cleaned and reconstructed TS images with a detail of Face.

3.3. Comparison of Images Wrapping Solids and Evaluation of a Sheet Configuration

Once obtained the cleaned body image of §3.2, it is then possible to use it for the acquisition of other information regarding if and how the TS man was really enveloped.

3.3.a. Verification of the TS body image wrapping

The frontal body image is 195 cm long while the dorsal one is 202 cm (dimensions taken before the 2002 intervention) and someone, perhaps forgetting the distortions due to wrapping, states that the TS could not have enveloped a man because the frontal image is much shorter than the dorsal one.

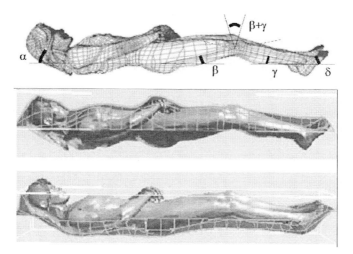

Figure 3.6. numerical manikin used to determine the TS enveloping and some the angular positions evaluated.

Basing on the detected anthropometric indexes (Fanti, Marinelli & Cagnazzo, 1999) evaluated in § 3.2b, it was built a numerical manikin and it was verified the possible correspondence of this manikin enveloped in a TS like sheet. This was done numerically by the superimposition of both the frontal and the dorsal body image on the manikin (see figure 3.6) (Basso & Fanti et al., 2001).

A full compatibility was detected if the TS Man is 175±2 cm tall[6] (see figure 3.6).

---

[6] It is interesting to observe that the compatibility was reached if a soft bed is supposed and if the position of the Man is not flat also confirming the presence of rigor mortis in the TS Man. The

3.3.b.   Evaluation of the configuration of the TS around the Man

Once defined the most probable shape and position of the body enveloped in the TS and basing on the hypothesis that the luminance levels of the image codify an information relative to the body-sheet distance, it is possible to reconstruct the configuration of the Ts around the TS Man.

Once fixed the position of the TS Man, the following simplified[7] relation was used to evaluate the body-cloth distance:

$$L = k/d + c \qquad\qquad (3.3)$$

where L is the luminance level of the body image, d the body-cloth distance and the constants k and c are experimentally evaluated from some reference points. The resulting TS enveloping on the body is reported in figure 3.7.

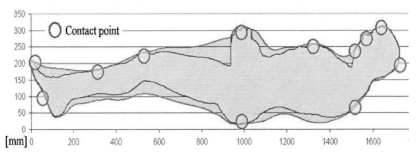

Figure 3.7. body-cloth distance evaluated from the luminance levels of the TS body image.

This results is important for the understanding of the body image formation process because it shows that many zones in correspondence of the dorsal image, especially legs, are not in contact with the enveloping cloth; as on the TS there are many scourge marks formed by body-cloth contact, it must be supposed that the blood stains were impressed on the TS before the body image formation and in a different configuration[8].

---

enveloped body probably kept the position it had had on the cross (apart from the arms, which were later folded over the pubic area), with feet stretched forwards and downwards (at angles δ of 34° and 30 ± 2°), legs partly bent (knee angles $\beta+\gamma$ of 19.5° and 23.5 ± 3°), and head falling forwards (α=30 ± 4°)

[7] The relation is simplified because for example it does not consider the point effect of an electrostatic field in the hypothesis of body image formation by means of corona discharge.

[8] Probably the soft bed composed of spices, flowers and plants caused a contact TS-corpse at the beginning, but after about 40 hours they got flabby, living some air between the corpse and the enveloping Cloth.

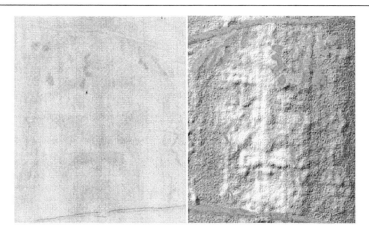

Figure 3.8. on the left contrast enhanced TS Face (courtesy B. Schwortz) to be compared with the 3-D specular reversion of the Face in which the position of the bloodstains on the hair, due to a gnomonic distortion, were corrected in correspondence of the cheeks.

3.3.c. Correction of the bloodstain position in the body image

The probable different enveloping configuration of the TS around the Body during the bloodstains transposition (before) and during the body image formation (after) detected in §3.3.b is also supported by the observation of some bloodstains in correspondence of the hair.

Even if one can suppose that the hair were soaked with blood, it is not easy to think that this hair could have generated almost the blood flows detectable on the TS image because these seem to better refer to a bleeding. Figure 3.8 reports in a 3-D processing the gnomonic correction of these stains that instead of on the hair they probably were located on the cheeks.

3.4. Luminance Analysis of Images

Luminance of images and their profiles can be used to compare and study the characteristics of photographs (Fanti & Moroni, 2002). In the case of the TS, this analysis was used to verify the compatibility of the TS Face digitized in a photograph with result of experiments. For example to determine the saturation degree, three different photographs of the TS were compared in terms of luminance levels with other two photographs relative to two experiments made to reproduce the TS body image. The first experiment involved the singeing of a cloth with a bas-relief as described by Pesce Delfino (Pesce Delfino, 2000) and the second involved the use a cloth with a bas-relief sprinkled with iron oxide powder containing traces of sulfuric acid as described by Nickell (Nickell, 1987).

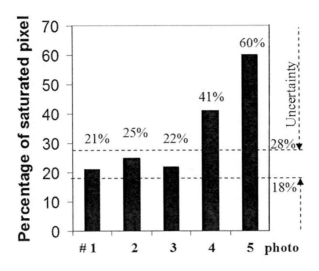

Figure 3.9. Saturation analysis by means of the "black" and "white" pixel percentage evaluation in some photographs of TS Face (#1, 2, 3) and experimental reproductions (#4 of Nickell and #5 of Delfino Pesce) (Fanti & Moroni, 2002).

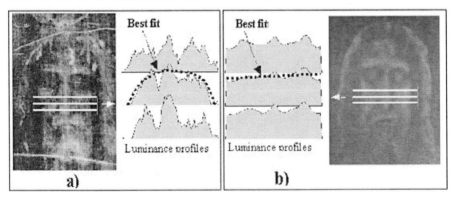

Figure 3.10. Processing of luminance levels of 3 horizontal lines (white) of the TS (a) and experimental results (b) obtained by scorching a sheet laid on a heated bas-relief. The 3D correlation with a cylindrical surface corresponding to a first approximation of the face wrapping is presumed in (a); instead a correlation with a slightly inclined plane corresponding to the bas-relief is evident in (b).

Each photograph digitized at 256 grey levels was analyzed in terms of luminance: the relative percentage of "black" pixels (values in the range from 0 to 5) and "white" pixels (values in the range from 250 to 255) was evaluated for each photograph. Figure 3.9 shows the interesting result that demonstrates the relative low level of saturation of the TS photographs (in the range between 18% and 28%) compared with the saturation levels of other experiments (41% and 60%): this again shows the difficulty to reproduce the TS body image in the details.

In another example (Fanti & Moroni, 2002) the results of singeing, presumed by Pesce Delfino (Pesce Delfino, 2000) as the cause of the body image formation, was compared in terms of luminance profiles with the TS Face, showing the inconsistency of this hypothesis. As shown in figure 3.10, if the luminance levels along a horizontal line are correlated with the body-sheet distance, the resulting surface of the TS Face appears almost cylindrical but the surface of the scorched sheet of Delfino Pesce appears almost flat. Perhaps the non-horizontal best fit of figure 3.10 b is caused by the fact that the sheet was first put on one side of the bas-relief.

3.5. Optical Filtering to Evidence the Double Superficiality of the Body Image

Optics can use powerful methods of image processing in order to enhance images characterized by signal-to-noise ratio lower than 1 (Fanti & Maggiolo, 2004). For example the photograph of the bs (back surface) of the TS in correspondence of Face and Hands has signal-to-noise ratio of about 0.5 because many disturbs practically cancel the body image to the naked eye; as it is shown in figure 3.11, some vertical bands also deteriorate the image.

There are different image processing methods but, in the hypothesis that the bands are perfectly vertical, a simple method consists in:

a. choosing a restricted area in which the image is not evident (such as the lower part of figure 3.11a);
b. evaluating the mean horizontal luminance profile of background (to reduce noise of white type);
c. constructing the compensation image that has dimensions equal to the initial image and a profile of luminance levels, in the horizontal direction, opposite to those of the mean background but constant in the vertical direction (plus a predefined constant luminance level) (figure 3.11c);

d. summing the initial image to the compensation one obtained in (c) to obtain the cleaned image (figure 3.11d) in which the background is more uniform and the body image is more evident.

A similar image processing could have been made using an FT (Fourier Transform) filter but in this case some detail of interest could be eliminated together with the noise. This method was used in the photograph of Face of figure 3.12 because there was no clear horizontal areas of only background without body image or local noise; the vertical band contained in the window of bs Face (figure 3.12a) was processed using FT algorithm to obtain the power spectrum (figure 3.12b) that was zoomed (figure 3.12c) to detect the spatial frequencies, evidenced by arrows, relative to the band to eliminate. The inverse FT transform of the power spectrum having the frequencies of interest reduced to zero (two black pixels of figure 3.12d) gives the cleaned result of figure 3.12e.

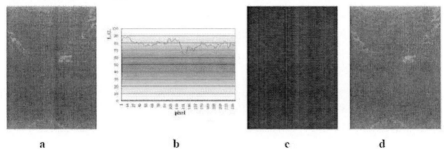

a   b   c   d

Figure 3.11. a) Image of bs hands showing a black vertical band; b) Mean luminance profile of (a); c) Compensation image; d) Resulting image of hands after summation of images of a) and c): black band is cancelled (Fanti & Maggiolo, courtesy of JOPA).

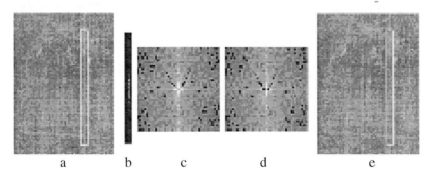

a   b   c   d   e

Figure 3.12. a) Vertical band windowed in Face image, bs; b) FT, power spectrum of window selected in a); c) magnification of b), corresponding to central area; d) two luminance levels (arrow), reset (black) to eliminate narrow segment; e) results of processing (Fanti & Maggiolo, courtesy of JOPA).

A similar image processing was employed to evidence the presence of a second body image in correspondence of Face and perhaps Hands on the TS bs (figures 3.13 and 3.14); the Face is very similar in shape and coincident in position to that of the frontal side. This presence was also numerically confirmed with the application of a template matching software using Vision 6.0® EasyAccess software that detected the correspondence of the nose-mouth area with a correlation coefficient of 86% when normally it is accepted the presence of the object if a coefficient of 65% or higher is reached.

The correlation formula used (Fanti & Maggiolo, 2004) to compute score S of a pattern at a given location is:

$$S = \frac{\sigma_{IP}}{\sqrt{\sigma_{II}\sigma_{PP}}}$$

(3.4)

where σ_{IP} is the covariance of the pixel value between the pattern and the part of the image used:

$$\sigma_{IP} = \frac{1}{nm}\sum_{i=1}^{n}\sum_{j=1}^{m}\left(I_{i,j} - m_I\right)\left(P_{i,j} - m_P\right)$$

(3.5)

σ_{PP} is the variance of pixel value in the pattern:

$$\sigma_{PP} = \frac{1}{nm}\sum_{i=1}^{n}\sum_{j=1}^{m}\left(P_{i,j} - m_P\right)^2$$

(3.6)

and σ_{II} is the variance of the pixel value in the part of the image used for correlation:

$$\sigma_{II} = \frac{1}{nm}\sum_{i=1}^{n}\sum_{j=1}^{m}\left(I_{i,j} - m_I\right)^2$$

(3.7)

$I_{i,j}$ is the pixel value at position (i, j) in the image relative to the pattern position tested, $P_{i,j}$ is the pixel value at position (i, j) in the scaled and rotated pattern, m_I is the mean pixel value in the part of the image used for correlation:

$$m_I = E(I_{i,j}) = \frac{1}{nm}\sum_{i=1}^{n}\sum_{j=1}^{m} I_{i,j}$$

$$(3.8)$$

and m_P is the mean pixel value in the pattern:

$$m_P = E(P_{i,j}) = \frac{1}{nm}\sum_{i=1}^{n}\sum_{j=1}^{m} P_{i,j}$$

$$(3.9)$$

The presence of a second body image on the bs coupled with the fact that it is very superficial, leaded to the conclusion that the body image is doubly superficial at least in correspondence of Face. This results sets some limits in the body image formation hypotheses discussed in Section 1 and privileges the hypotheses based on corona discharge mechanisms because they typically generate doubly superficial images (Fanti, Lattarulo & Scheuermann, 2005).

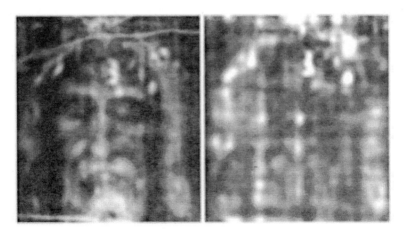

Figure 3.13. processed images of Face: frontal side (on the left) and bs (on the right), showing similar body features (Fanti & Maggiolo, courtesy of JOPA).

Other techniques based on the frequency filtering using the FT can be employed to clean in other ways the photograph of the TS from the typical horizontal and vertical bands, but having those bands not constant spatial frequencies, the computer cleaning can be not perfect and some information relative to the body image can be lost; an example is reported in figure 3.15.

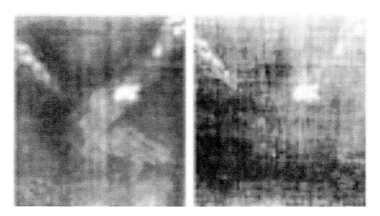

Figure 3.14. Processed images of arms and hands, frontal side (left) and bs (right). Low resolution of bs image does not show fingers, but a clearer area in center, surrounded by a darker area, may be imprint of hands (Fanti & Maggiolo, courtesy of JOPA).

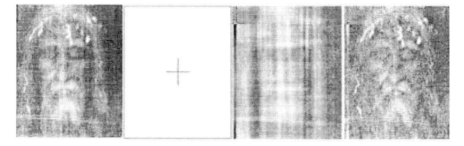

Figure 3.15. the banded image of Face on the left was cleaned (image on the right) subtracting the result of a FT of a cross filter shown in the center with the part of the image subtracted (Doumax & Porter).

3.6. MTF Technique Applied to the TS

The MTF (Modulation Transfer Function), frequently used to evaluate the characteristics of the photographic objectives, was applied (Fanti, Sept. 2005) to quantify the resolution of the of the TS body image empirically evaluated in a previous paper (figure 3.16) (Fanti & Maggiolo, 2004) by means of the back-forward FT in which it resulted that the TS body features are characterized by frequencies greater than 80 [m^{-1}] but less than 100 [m^{-1}], i.e. details with a spatial period of about 10 mm and therefore a resolution of about 5 mm.

Figure 3.16. resulting inverse FT of image of Face, in which all frequencies lower than 62.1[m-1] are eliminated with many body features except those of lips (on the left); the other results after high-pass filtering of spatial frequencies lower than 81.2 [m-1] and 100.4 [m-1] show (respectively on the center and on the right) that body features are also cancelled (Fanti & Maggiolo, courtesy of JOPA).

The system transfer function can be obtained through the EET (Edge Exposure Technique) by processing the image generated by an optic system that acquires a step function named ESF (Edge Spread Function). By means of the first derivative the LSF (Line Spread Function) is obtained and its FT furnishes the MTF, see figure 3.17.

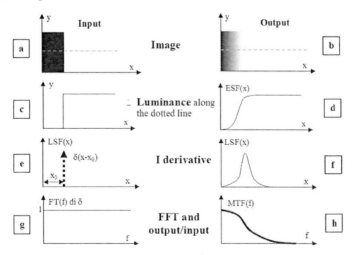

Figure 3.17. Edge Exposure phases. -1) A step function image is acquired (a) obtaining as output the response of the optical system (b); -2) The luminance levels along the dotted lines are evaluated (c, d) to obtain the ESF. -3) The first spatial derivatives are then calculated (e, f) and the LSF is obtained (the first derivative of the Heaviside function is the Dirac function). -4) As the FT of the Dirac function gives a flat spectrum (g), the optical system MTF (h) is the FT of the LSF (f).

The MTF of the TS images was then evaluated with the corresponding higher uncertainty due to the approximated method, evaluating the ratio between the FT of the TS image (output) and the averaged FT of the corresponding anatomical details (input) according to the formulas:

$$H(f_x,f_y) = \frac{G(f_x,f_y)}{F(f_x,f_y)} = MTF(f_x,f_y) \cdot e^{i\varphi(f_x,f_y)}$$

(3.10)

$$MTF(f_x,f_y) = \left| \frac{G(f_x,f_y)}{F(f_x,f_y)} \right|$$

(3.11)

where $H(f_x, f_y)$, also called OTF (Optical Transfer Function), $G(f_x, f_y)$ and $F(f_x, f_y)$ are respectively the FT of $h(x,y)$, $g(x,y)$ and $f(x,y)$; $h(x,y)$ is defined as the ratio between the output $g(x,y)$ and the input $f(x,y)$; f_x, f_y are the spatial frequencies along the directions x e y.

In the case of hands, the respective 2D-FT are shown in figure 3.18 where it can be observed the prevalent information along the direction of 58° corresponding to that of the fingers that are inclined of -32°. The luminance levels profiles of figure 3.18b,d along the direction of 58° are reported in figure 3.19a; the MTF curve along the same direction is reported in figure 3.19b.

Assigned a minimum uncertainty of ±0,1 to the MTF curve, the characteristic frequencies are: $f^* = 63 \pm 5$ m^{-1} corresponding to $MTF(f^*)=0,5$; $f^{**} = 104 \pm 10$ m^{-1} corresponding to $MTF(f^{**})=0,05$ that is compatible with that of Face ($=101 \pm 9$ m^{-1}).

After the comparison with other results of Hands and Face the following results can be assigned for the characteristics of the TS body image:

$$f^* = 60 \pm 10 \ m^{-1} \text{ corresponding to } MTF(f^*)=0.5;$$

$$f^{**} = 103 \pm 10 \ m^{-1} \text{ corresponding to } MTF(f^*)=0.05.$$

Observed that in the case of a frequency of 100 m^{-1} a couple of b-w lines having a dimension of 10 mm are visible, it can be therefore concluded that the maximum resolution of the TS body image evaluated at an MTF value of 0.05 (contrast of 5%) is 4.9±0.5 mm in agreement with what was empirically evaluated in a previous paper (Fanti & Maggiolo, 2004).

It must be noted that the MTF curves of the TS images show a behaviour different from those typical (figure 3.17 h) because an anomalous peak at $f_p=32$ m^{-1} is present. It therefore appears that the spatial frequencies lower than 20 m^{-1}

(lines about 25 mm thick) are not well defined in the TS images but those around 30 m^{-1} (lines about 15 mm thick) are better defined: future studies can confirm this fact perhaps useful to characterize the body image formation mechanism.

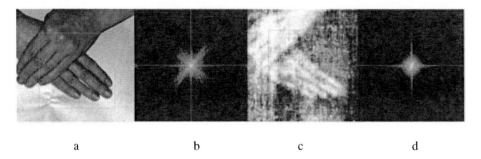

a   b   c   d

Figure 3.18. a) input image of hands; b) 2D-FT of image(a); c) output image of TS hands; d) 2D-FT of image (c).

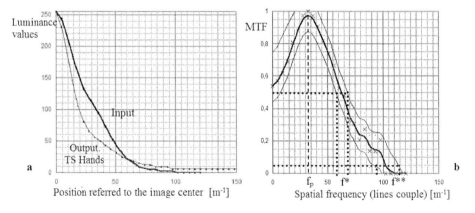

a   b

Figure 3.19. -a) equalized luminance profiles of images of figure 3.18(b, d) along the direction of 58°; -b) MTF curve of the TS hand: the frequency peak at fp=32 m-1 is probably typical of the TS image.

3.7. Detail Characterization by Means of Color Analysis

RGB values of a digital photograph are data not so accurate as those obtainable by means of a spectrophotometer, but in the lack of these, they can be used to extract some color information. The ratios *R/G* or *R/B* or the evaluation of the corresponding vector *L* in the *R-G-B* space evaluated with equation (2.1) can be used to evidence the difference of color in some details, even if this difference is not so relevant as in the TS case.

This technique was applied by Propp and Jackson (Propp & Jackson, 1997) without particular results of interest because "precision data is required for clarification of this potentially important results" and it has been applied by the authors of the present work to the new photograph of the TS made in 2002 (Ghiberti, 2002) to evidence for example the presence of different kinds of blood on the TS.

Someone (Jackson, 1998) states that the OTB (OuT-of-image Blood), such as that in correspondence of the elbow (see figure 3.20) has a color slightly different from that of the OB (On-image Blood) also explaining the difference with the radiation received during the body image formation.

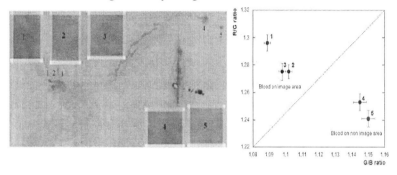

Figure 3.20. OB and OTB color analysis by means of the evaluation of the G/G and G/B ratios; on the left 3 image spots of OB relative to the wrist wound were compared with 2 image spots of OTB taken near the eloow; on the right, the corresponding G/G and G/B ratios plot shows that the OB color is really different from that of OTB.

The difference between OTB and OB color is confirmed by the R/G – G/B ratios plot of figure 3.20 in which the Spots #1, 2, 3 (respectively of 9x12, 10x13 and 12x12 pixels) taken from the OB wrist wound are clearly in a different position in the plot from Spots #4, 5 (respectively of 12x9 and 11x10 pixels) taken from the OTB near the elbow; Table 3.2 shows the relative results in terms of R/G, R/B and G/B ratios with their uncertainty values.

Table 3.2. R/G, R/B and G/B ratios with their uncertainty values evaluated at 68% confidence level of the OB and OTB spots reported in figure 3.19

R/G	unc.	R/B	unc.	G/B	unc.
1	+1.296	+0.006	+1.416	+1.089	+0.002
2	+1.275	+0.005	+1.408	+1.102	+0.002
3	+1.275	+0.006	+1.406	+1.098	+0.002
4	+1.253	+0.006	+1.447	+1.145	+0.004
5	+1.241	+0.006	+1.438	+1.150	+0.004

3.8. Comparison of Images Acquired with Different Light Spectra

It is well known that, depending on the characteristic of incident light spectra, the reflected image can be different if it is very superficial; this technique is frequently used in IR light for example to examine the presence of drafts under paintings. In the case of the TS the image is very superficial because it was probably generated by a flash of energy, the image formation did not go to completion and that its intensity depends principally from an areal distribution of almost equally colored linen fibers placed side by side with non image fibers (Fanti, 24 authors, 2005). It can be hypothesized that the TS body image has not only an areal but also a volumetric distribution in the sense that the depth of the image is higher in correspondence to the zones where the energy intensity, and than the color, was higher.

To test this hypothesis some photographs of Face made with different spectra of incident light can be compared. This study, still in progress, reports the first results relative to the comparison of photographs made in IR, Red, Green, Blue and UV light. Figure 3.21 shows the results of averaged luminance profiles relative to an horizontal line passing through the nose when the Face is illuminated by means of IR, Green and UV light.

A part from some noise still present, it appears that in correspondence of Face the IR image shows peaks more evident than images acquired in IR and Green light. It is known that in general, due to their different wavelengths, the IR light goes more in depth into the fabric than the UV light. The IR light gives then information relative to a thicker layer of the fabric and in the present case it seems to show that the Face image vanishes where the intensity is lower (for example in the area between the nose and the cheeks); inversely the UV image that acquire a more superficial layer, seems to have a more flattened luminance profile confirming that a thinner layer of TS color is more distributed on the fabric surface but a thicker layer can only be found where the body image is more evident (for example the tip of the nose). This is a preliminary confirmation that the TS body image color is not characterized by a constant thickness, but it is proportional to the image luminance and then to the energy intensity that produced it.

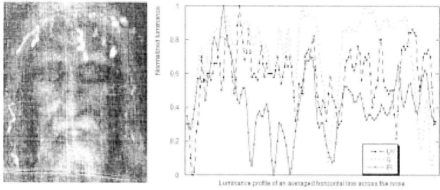

Figure 3.21. averaged and normalized luminance profiles of the pink horizontal line crossing the nose seen in IR, Green and UV light.

3.9. Images Overlapping: Optical Distortions and Detail Analysis

By means of computer analysis of images it is possible overlap and compare photographs of deformed shapes as it was discussed in §3.3. This technique can be employed to directly measure the distortions caused by a cloth wrapping a body or to have information about a cloth folding.

  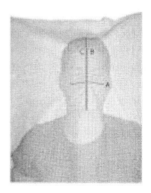

Figure 3.22. by means of the overlapping of a photograph of face with a photograph of a face enveloped on a sheet, it was possible (Courtesy of Latendresse, 2005) to compare the real length of the two segments (shown on the right) with their distorted length in the photograph; horizontal and vertical segments has respectively lengthening ratio in the sheet equal to 12/11.3=1.06 and 23.3/21.5=1.08: the distortion is less than 10%.

M. Latendresse (Latendresse, 2005) showed the incorrectness of the argument proposed by some researcher against the TS authenticity: they are convinced that, to avoid prominent image distortions, the top half of the Shroud could not have been draping a real body when the images formed, in fact an highly distorted body image should have resulted.

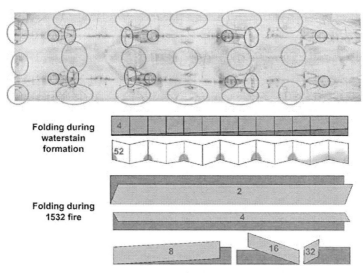

Figure 3.23. waterstains (green) and fire marks (red) on the TS (upper image) used to determine the different folding (courtesy of A. Guerreschi & M . Salcito, 2002).

Latendresse showed that if a cloth is appropriately laid on the front part of a body, "no major image distortions occur even if small image distortions are to be expected" as can be observed on the TS. This fact derived from length measurements on digital images of distorted shapes showing that (see figure 3.22) if a sheet partially covers a face, the corresponding contact body-sheet area is distorted of less than 10%.

Another application was proposed by A. Guerreschi and M. Salcito (Guerreschi & Salcito, 2002) who superimposed all the supposed rectangles resulting from the TS folded in different ways. They detected the folding compatibility with the burn marks and the waterstains present on the TS, but they demonstrated that the TS was folded in a different way during the waterstains forming and the damage of 1532 Cambéry fire, see figure 3.23.

3.10. Optical Comparison of Computer Generated Images

Many scientists qualitatively state that the TS body image was obtained by an highly directional source of energy. To quantify this characteristic a work of Fanti

(May 2005) simulated the effect of an hypothetical radiation emitted by parts of human body enveloped by a sheet to better understand the mechanism that could have generated the body image.

Computer software based on the Lambert law used finite element meshes of parts of human bodies such as hands and face to evaluate the radiation intensity incident on all the elemental areas of an enveloping numerical sheet. In the hypothesis of proportionality between radiation intensity and image luminance, different pictures were obtained that allowed a comparison with the TS body image.

The following equations for Q, total radiative power exchanged between two surfaces (human skin and sheet), in reference to m couples of elemental surfaces, were used to simulate the effects of different kinds of radiation:

$$Q = \sum_{i=1}^{m} dq_{i(1-2)} \quad ; \quad (3.12)$$

$$dq_{i(1-2)} = \varepsilon_\varphi \cdot dA_i \cos\varphi_i \frac{dA_2 \cos\varphi_2}{r^2} \quad (3.13)$$

where ε_φ is the directional emissivity of a non-metallic material, dq_i is the power exchanged between two elemental surfaces, dA the elemental area of each surface element and ϕ the angle defined by the normal directions of the two surface element considered.

The different types of radiation considered were (see figure 3.24):

a) perfectly diffusive radiation defined by Lambert law ($\varepsilon_\phi = 1$),
b) radiation only parallel to gravity acceleration,
c) radiation only normal to the skin ($\varepsilon_{\varphi1=90°}=1$; $\varepsilon_{\phi1\neq90°}=0$).

The numerical hand results obtained by this simulation are reported in figure 3.25.

It is qualitatively evident that the radiation normal to the skin[9] of figure 3.25c is the best approximation of the TS Hands. This result is also quantitatively confirmed evaluating the luminance profile of each image; figure 3.26 shows the luminance profile relative to TS hand compared with the image of the hand in the case of radiation normal to the skin.

---

[9] In a first approximation the radiation normal to the skin is explainable if a corona discharge hypothesis of image formation is assumed [Fanti, Lattarulo & Scheuermann].

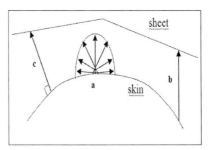

Figure 3.24. mesh of hand used to simulate radiation (on the left) and different kinds of simulated radiative exchanges (on the right).

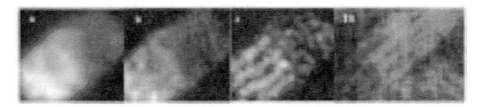

Figure 3.25. numerical hand images resulting in the cases corresponding to figure 3.24 (-a) diffusive radiation, -b) radiation parallel to gravity acceleration, -c) radiation normal to the skin) compared with the TS Hands.

The template matching applied to each luminance profile of the images shown in figure 3.25 has been evaluated by means of equations similar to eq. (3.4). The best matching was obtained in the case of a uni-directional emissivity normal to the skin surface, with a correlation coefficient R=0.93 in reference to the TS hand. This leads to suppose that the radiation that generated the TS image was highly directional and normal to the skin.

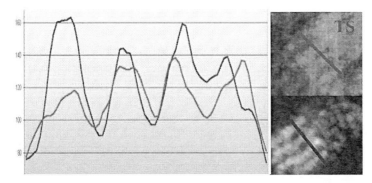

Figure 3.26. luminance profiles along the indicated lines, relative to the TS Hand photograph (red line) and the simulated image of hand in the case of radiation normal to the skink (of figure 3.24) (Fanti, may 2005).

3.11. Discussion

In reference to the TS general characteristics discussed at Section 1.3 and to the Optical characteristics of Section 1.4 it is evident the very important contribution of optics in the research of the TS that allowed to detect the very particular characteristics of the TS body image that up to now they are not yet reproducible all together.

In particular it is to evidence that all the points O1-O9 of Section 1.4 derive from optical observations relative to the studies reported in Section 2 (Past) or Section 3 (present) and that also many characteristics of Section 1.3 still derive from the important optical studies done up to now.

Not only from a strictly optical point of view the research on the TS is obviously still open and perhaps it is known only the 5% of the whole information relative to the most important Relic of the Christianity; even if we know that the body image probably formed after a brief but intense energy emission perhaps connected with a corona discharge, many details and many proofs are still unknown and for this reason Optical Science in particular and Science in general have still to do many important steps using more recent or new techniques.

For example Raman spectroscopy can be used in the chemical identification of some components in the image fibers, since vibrational information is very specific for the chemical bonds in molecules. Raman scattering by a crystal gives information on the crystal orientation and the polarization of the Raman scattered light with respect to the crystal can be used to detect the crystal structure of the polyshaccarides thin layer around the image fibers or of the cellulose in the image fibers, giving useful information about the mechanism of the body image formation. The same Raman techique can also be used, coupled with cross-polarization in the visible light, to have information about possible shift in Raman frequency with applied stresses.

4. OPTICAL RESEARCH: THE FUTURE

4.1. What Has It Been Done and What Will It Be Done?

Someone may ask: has all the possible contribution of optics research applied to the TS been exhausted? If not, what it can be done to know new facts about the most important Relic of the Christianity? Certainly only a minimal part of the TS secrets has already been discovered and the optics has always played a very important role in the research.

Even if the beginning of the third millennium is a period in which the TS owners are not very open to new researches, many studies are still in progress. For example in 2002 a group of researchers interested in scientific aspects relative to the TS formed a no profit **Internet Group named ShroudScience** that on "Yahoo!" discusses new scientific studies and their results; at the end of 2006 this Group was composed of about 100 members and more than 9000 e-mails were exchanged (every researcher interested in a possible participation can contact the first author of this paper).

The Group, very interested in better understanding how the TS body image was formed, has published in 2005 a list of evidences to be considered for testing hypotheses about the TS image formation hypotheses (Fanti, 24 authors, 2005); in the next future it is foreseen that every scientific hypothesis will be discussed against the evidences reported in that list, in order to detect which is the most reliable.

From a more strictly optical point of view many researches can be proposed when the Turin authorities will be more open to new studies, but for the moment some proposal have still been compiled. For example, in the occasion of World Congress on the TS of Orvieto (Italy) 2000, a proposal (Basso, Fanti, Marinelli et al., 2000) for the construction of a big Calibrated Multiresolution Atlas of the TS was compiled in order to fill some voids in the optical data about the TS.

The construction of a big Atlas is very important for future studies because it should contain a very detailed data base for future researches and it should be accessible from any scientist interested in the TS research. It could be made in different forms: as a book containing a great variety of photos, from the whole image to microscopic details; as a digital atlas put into Internet for all the persons interested in general studies; as a complete database codified in digital form (DVD) containing all the calibrated high resolution images acquired. These data will be used by researchers that will go more in depth in reference to some particular aspects regarding the TS. The proposal is subdivided in the following sections:

- Design and construction of 2-axes manipulators;
- Hardware and data acquisition;
- Visible, IR, UV lighting, shaving light included;
- Calibration procedures and uncertainty analysis;
- High resolution digital photography;
- High resolution analog photography;
- CCD colorimetry;
- Microphotography of body image in situ;

- Mid -IR, Near-IR, VIS and Near-UV spectroscopic measurements;
- Microscopic FT-IR examination;
- Mapping with a scanner also in UV light;
- Drawing some fibers;
- Partial mapping of the back of the Shroud also with endoscopy in visible light, UV and IR light;
- Collection of dusts and fine materials to be successively photographed by means of microscopes.

Also in reference to a previous paper (De Cecco & Fanti, 1998) the Atlas propose an innovative method for the TS image acquisition based on an integration sphere that makes the TS albedo independent on the illumination direction, see figure 4.1. It is in fact well known that the intensity of body image on the TS is dependent on the illuminant inclination. This technique would then allow us to acquire quantitative photographs of the TS.

In 2002 some partial work has been done but it is surely a minimal part with respect to what was proposed in the Atlas; from 2000 up to now other new optical facilities are at disposition, so some improvement must be foreseen before to start, but this project still remains of fundamental importance.

It has been discussed what it has been detected on the TS and what it is not yet done; here are discussed some possible improvements to the optical research on the TS. For example in Section 3.9 "*Comparison of image acquired with different light spectra*" it has been showed that probably the body image thickness is not constant, but dependent on the energy intensity that generated the image; to have a definitive answer a series of calibrated images acquired from the UV to IR range is necessary. Furthermore the thermal diffusion measurement as a function of time, acquiring IR photographs of the TS posed in front of a uniformly heated body, could be an aid to confirm this hypothesis.

In Section 3.3 "*Comparison of images wrapping solids and evaluation of a sheet configuration*" and 3.10 "*Images overlapping: Optical distortions and detail analysis*" some answers were done about the corpse enveloped in the TS, but future studies based on a more detailed comparison between a computer manikin and the corresponding image obtainable on a sheet must be performed also as a function of the selected body image formation hypothesis; for example in the corona discharge hypothesis, also the image intensity variation due to the point effect must be accounted for. In this view the models discussed in Section 3.11 "*Optical comparison of computer generated images*" can be improved using a mesh corresponding to a complete human body.

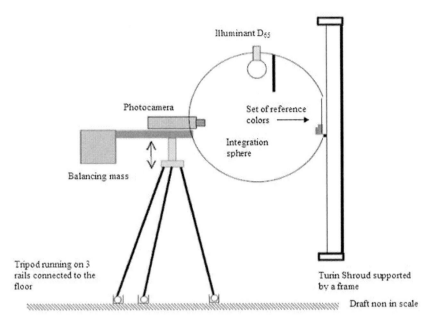

Figure 4.1. Scheme (De Cecco & Fanti, 1998) of an acquisition system for the TS based on an integration sphere to avoid the effect of the illuminant direction.

The "*Detail characterization by means of color analysis*" discussed in Section 3.8 will be more accurate if based on quantitative photographs that can be obtained for example by means of the system proposed in figure 4.1. Also techniques like those discussed in Section 3.6 "*The MTF technique applied to the TS*" should be improved to characterize the body image in terms of spatial frequencies; this will also help to find a reliable hypothesis of body image formation.

The TS is highly damaged due to the 2000 years of history and a computer cleaning should be necessary from a point of view of the body image analysis; this has been partially done in Section 3.2 "*Computerized body image cleaning reconstruction*" but something more accurate also based of the newer photographs is surely useful for a detailed analysis also in reference to the body image formation. Furthermore the typical horizontal and vertical bands also due to the folding should be eliminated by means of the techniques discussed in Section 3.5 "*Optical filtering …*" or with the employment of other newer techniques such as the 2-D wavelet analysis.

Wavelet technique seem very promising for computer the cleaning of TS bands because these bands are not spatially periodic on the fabric[10]. The Short Time Fourier Transform or better the Wavelet transform is therefore more suitable than the traditional Fourier transform that bases itself on the hypothesis that the signal are spatially periodic.

As a first example figure 4.4 shows the preliminary result of the application of a 2-D wavelet transform done in correspondence to the TS Hands to reduce the vertical bands on the image. This promising results shows that some noise can be removed without reducing the characteristics of the body image. Opposite to the merit of this technique that works locally on the image, with the noise reduction frequently the Fourier transform also reduces the information of the image of interest, see for example figure 3.14).

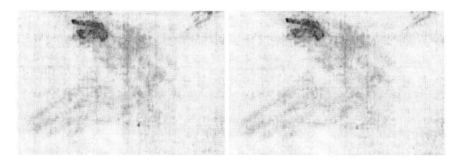

Figure 4.2. image of TS Hands (on the left) and partially cleaned (on the right) using a 2-D wavelet analysis.

In addition, the future optical studies should also include an improved analysis of the test done up to now, for example it must be foreseen an high resolution acquisition of body image in the IR spectrum ranging from 1 to 14 micrometers (see Section 2.6d *"Thermographic investigation"*) and some experimental results of images impressed in linen fabrics subjected to different sources of energies must be studied and compared with the TS image. In this view some techniques based on excimer lasers (Baldacchini et al., 2006) could be employed to reproduce the TS Face or other anatomical parts. The energies and the times required to reproduce TS-like images will be useful to understand which process could have caused the body image on the TS.

Obviously with the time passing many optical technologies will be improved and always better results will be available if new instruments will be applied to

---

[10] The irregularities of the bands are due both because TS was hand-weaved with linen coming from different bleaching-baths and because it was folded then generating crimps of different types.

these studies; for example the resolution of the images will be certainly improved and these images coupled with the higher computer capacity will allow more accurate image processing to better detect the details present on the Relic.

4.2. Optics for the TS Conservation

In Section 1.3 some peculiar characteristics of the TS are listed and in particular G3, G6, G7, G8 and G13 show that the body image can be seen because of a precocious ageing of the dehydrated polysaccharides thin layer of the image fibers. This means that with the time passing, the clearer background of the TS darkens becoming similar to the image, then cancelling the bodily imprint on the linen fabric.

To reduce this effect the TS is conserved in a dark container at uniform temperature and rarely can it be exposed. For example some researchers that had the occasion to see the TS during both the 1978 and 2000 exposition and during the private exposition of 2002 had subjectively noticed the lack of contrast of the image exposed in 2002. Therefore from a scientific point of view, in agreement with what it is discussed in Section 4.1, many new test should be done on the TS to unveil many points still obscure, but from a conservation point of view it is necessary to perform a reduced number of test, mainly avoiding those who require high-power light sources, in order to reduce the ageing process that tends to cancel the body image; the exposition to intense light sources for hours and days during the 2002 intervention surely acted in some way reducing the image contrast with background.

The reduction of the ageing effects such as light and temperature acting on the TS is accepted by all the scientists from a qualitative point of view, but very few is known from a quantitative point of view. Therefore before to decide if a test will be allowable, some conservation test must be designed and done on samples similar to the TS image exposed to known light sources for predefined time durations. The comparison among the luminance level of the photographs of these samples made during various steps of illumination will allow some quantitative evaluation about the effective influence of light on the TS image degradation.

On the other hand it is necessary to have quantitative photographs of the TS, also proposed in the Atlas project (Basso, Fanti, Marinelli et al., 2000), that will be used in the future as comparison if the contrast degradation of the image versus time will be monitored. For this purpose an acquisition system like that proposed in figure 4.2 can be employed, but the set of reference colors should include colors relatively easy to produce such as plates of pure gold and platinum having a well known surface finish such as lapping. In this way the reference color reproducibility will be easier to obtain also in the future centuries when probably different materials will be employed for optical purposes.

CONCLUSION

This paper has shown the important contribution of optics to the detection of the peculiar characteristics of the TS and in particular of the double body image impressed on it.

In the first Section these particular characteristics are discussed showing that up to now the Science and the Technology is not able to reproduce an image having such characteristics so the main problem in reference to the TS is that to discover how the body image was formed.

Many studies have been done also in this view and they are described in Section 2 (*The Past*) and in Section 3 (*The Present*) but still many experiments must be done in the future and some of them are presented in Section 4 (*The Future*).

In the view of solving the problem regarding the understanding of how the body image was formed, the optics research will surely play a very important role in the future TS research. In any case the complexity of these studies implies that the future analyses will not be let to a limited group as it was done in the very first years of the third millennium, but to a wide commission of experts able to well represent each of the many disciplines involved in the TS studies; it will also be better if the future wide commission will be leaded by a scientific organization without any religious preconception.

The future studies, if allowed by the Vatican and by the Turin Authorities, performed not neglecting the conservation problems, will probably allow us to disclose some secrets of the TS that allow to certify it as the most important Relic of the Christianity.

Up to now it is known that the TS enveloped a corpse that did not let putrefaction signs because it was removed not later than 40 hours. The hardly wounded copse did not let any smear in correspondence of the blood stains letting to think that it leaved the linen fabric in a particular way very difficult to reproduce. The image that that corpse impressed on the TS is not explainable by Science and Technology and a probable hypothesis of image formation is referred to a source of energy that acted at a distance for a short time interval (perhaps some microseconds) perhaps connected to the corona discharge generated by an intense electrostatic field. (also of some millions of volts).

From the point of view of a positivistic science the TS body image should not exist but it is an Object that everyone can admire and measure or directly of by means of photographs; therefore the actual Science must admit its limits and to explain this Object it must go out of Science.

From a metaphysical point of view it can be said that many signs and the Christian tradition join the name of the TS Man with that of Jesus of Nazareth and what can be detected on the TS is in agreement with the Holy Scriptures, but there is also reported that Jesus resurrected. As for the moment the body image formation mechanism is not scientifically explained, the simplest solution to the problem is not that to think that the body image formed due to an energy source that was perhaps a by-product of the Resurrection?

REFERENCES

[1]   Accetta, J. S. and Baumgart, J. S. (1980). Infrared reflectance spectroscopy and thermographic investigations of the Shroud of Turin. *Applied Optics, 19*, 1921-1929.
[2]   Adler, A. D. (1996). Updating recent studies on the Shroud of Turin. *Archaeological Chemistry, 625*, 223-228.
[3]   Adler, A. D. (2002). *The Orphanated manuscript*. A Shroud Spectrum International Special Issue, Torino, Italy: Effatà ed. Cantalupa.
[4]   Allen, N. P. L. (1993). Is the Shroud of Turin the first recorded photograph? *South African Journal of Art History, 11*, 23-32.
[5]   Archdiocese of Turin (2000). Sindone le immagini 2000 Shroud images. Torino, Italy: ODPF.
[6]   Artom, M. and Soardo, P. (1981). *Caratteristiche fotometriche e colorimetriche della S. Sindone. In La Sindone Scienza e Fede*, pg. 321-329, Bologna 1981, CLEUB Bologna, Italy.
[7]   Ashe, G. (1966). What Sort of Picture. *Sindon,*15-19. Torino, Italy.
[8]   Basso, R., Bianchini, G. and Fanti, G. (2000). Compatibilità fra immagine corporea digitalizzata e un manichino antropomorfo computerizzato. *Congresso Mondiale "Sindone 2000"*, Orvieto, 27-29 Agosto 2000, www.dim.unipd.it/fanti/Manikin.pdf
[9]   Basso, R., Fanti, G., Marinelli, E. and et al. (2000). Proposal for the Construction of a Calibrated Multiresolution Atlas of the Turin Shroud. *http://web.tiscali.it/sindone 2000/ATLAS.PDF*

[10] Baldacchini, G., Fanti, G. and et al. (2006). Colouration of Linens by Excimer Lasers and Comparison with the Body Image of the Turin Shroud. (in Italian, summary in English language) *Internal report of ENEA,* Frascati, Rome, Italy: Frascati Research Center.

[11] Craig, E. A. and Bresee, R. R. (1994). Image Formation and the Shroud of Turin. *Journal of Imaging Science and Technology, 38,* 1, 59-67.

[12] Cunico, A. (1999-2000). Ricostruzione dell'immagine corporea dell'Uomo della Sindone: analisi preliminare della correlazione tra luminanza e distanza corpo-telo. *Degree thesis tutored by G. Fanti, academic year* 1999-2000, Padua University, Italy: Dipartimento di Ingegneria Meccanica.

[13] Damon, P. E. and et al. (1989). Radiocarbon dating of the Shroud of Turin. *Nature, 337,* 611-615.

[14] De Cecco, M. and Fanti, G. (1998). Studio di un sistema di visione per la mappatura colorimetrica della Sindone di Torino. *III Congresso Internazionale di Studi sulla Sindone.* Torino, Giugno.

[15] De Gail, P. (1972). Le visage de Jesus Christ et son linceul. Paris: Éditions France-Empire.

[16] De Liso, G. (2002). Verifica Sperimentale della Formazione di Immagini su Teli Trattati con Aloe e Mirra in Concomitanza di Sismi. IV *Int. Scientific Symposium on the Turin Shroud*, Paris, 25-26 April 2002,

[17] Doumax, R. and Porter, D. www.shroudstory.com/Filtered/front-filtered-neg-histo-g1-final.jpg, www.shroudstory.com/Filtered/front-source-histo-g1-final.jpg, www.shroudstory.com/Fil tered/d-hen-9-260.jpg

[18] Enrie, G. (1933). *La Santa Sindone.* Torino, Italy: Società editrice Internazionale.

[19] Ercoline, W. R., Downs, R. C. Jr. and Jackson, J. P. (1982). Examination of the Turin Shroud for image distortions. *IEEE 1982 Proceedings of the International Conference on Cybernetics and Society, October* 1982, pp. 576-579.

[20] Fanti, G., Marinelli, E. and Cagnazzo, A. (1999). Computerized anthropometric analysis of the Man of the Turin Shroud. *Int. Research Conference "Multi disciplinary Investigation of an Enigma,"* Richmond, Virginia, U.S.A., 18-20 June 1999, internet: www.shroud.com/pdfs/marineli.pdf

[21] Fanti, G., Marinelli, E. (2000). *Cento prove sulla Sindone: un giudizio probabilistico sull'autenticità.* Padova, Italy: Messaggero di S. Antonio.

[22] Fanti, G. and Faraon, S. (2000). Pulizia e ricostruzione computerizzata dell'immagine corporea dell'Uomo della Sindone. *Congresso Mondiale "Sindone 2000",* Orvieto, 27-29 Agosto.

[23] Fanti, G. and Marinelli, E. (2001). A study of the front and back body enveloping based on 3D information. *Dallas International Conference on the Shroud of Turin,* Dallas, Texas, U.S.A., 25-28 October, 2001, www.dim.unipd.it/fanti/ Enveloping.pdf

[24] Fanti, G. and Moroni, M. (2002). Comparison of luminance between **the** face of the Turin Shroud Man and experimental results. *Journal of Imaging Science and Technology, 46,* 2, 142-154. Internet: http://www.imaging.org/store/epub.cfm? abstrid=8125

[25] Fanti, G. and Maggiolo, R. (2004). The double superficiality of the frontal image of the Turin Shroud. *Journal of Optics A: Pure and Applied Optics,* 6, 6, 491503, www.sindone.info/ FANTI.PDF

[26] Fanti, G. and et al. (24 authors) (2005). Evidences for Testing Hypotheses about the Body Image Formation of The Turin Shroud. *The Third Dallas International Conference on the Shroud of Turin,* Dallas, Texas, September 8-11, 2005, http://www.shroud.com/ pdfs/doclist.pdf

[27] Fanti, G., Lattarulo, F. and Scheuermann, O. (2005). Body Image Formation Hypotheses Based on Corona Discharge. *The Third Dallas International Conference on the Shroud of Turin,* Dallas, Texas, September 8-11, 2005, *http://www.dim.unipd.it/fanti/corona.pdf*

[28] Fanti, G. (2005). Numerical Analysis of the Mutual Radiation Effects of Complex Surfaces. *2nd Italy – Canada Workshop on: 3D Digital Imaging and Modelling – Applications of Heritage, Industry, Medicine and Land,* Padova, May 17-18 2005, www.dim.unipd.it/ fanti/rad-skin.pdf

[29] Fanti, G. (2005). Valutazione della risoluzione di immagini mediante analisi del modulo della funzione di trasferimento. *VI Congresso Nazionale di Misure Meccaniche e Termiche,* Desenzano del Garda, 12 14 Settembre 2005, http://archimedes.ing.unibs.it/MMT/ articoli/10-Risoluzione_ Immagini.pdf

[30] Filas, F. (1982). The Dating of the Shroud of Turin from coins of Pontius Pilate. Cogan, Youngtown, Arizona, USA.

[31] Frei, M. (1979). Il passato della Sindone alla luce della palinologia. In: *"La Sindone e la Scienza", Atti del II Congresso Internazionale di Sindonologia, Torino* 1978, 191-200. Torino: Edizioni Paoline.

[32] Gastineau, P. (1986). In *Shroud Spectrum International: "A Bas relief from a Photograph of the Holy Face by Paul Gastineau",* Issue #18, March 1986 www.shroud.com/spec trum.htm.

[33] German, J. D. (1977). An Electronic Technique for Constructing an Accurate Three-Dimensional Shroud Image. *Proc. of the 1977 United States*

Conference of research on the Turin Shroud, Albuquerque, New Mexico, USA, March 1977.

[34] Ghiberti, G. (2002). *Sindone le immagini 2002 Shroud images.* Torino, Italy: ODPF.

[35] Gilbert, R. Jr. and Gilbert, M. (1980). Ultraviolet-visible reflectance and fluorescence spectra of the Shroud of Turin. *Applied Optics, 19*, 1930-1936.

[36] Guerreschi, A. The Turin Shroud: from the photo to the three-dimensional. www.shroud.com/pdfs/aldo1.pdf and "The Turin Shroud and Photo-Relief Technique", http://www.shroud.com/pdfs/aldo2.pdf

[37] Guerreschi, A. and Salcito, M. (2002). Photographic and Computer Studies Concerning the Burn and the Water Stains Visible on the Shroud and their Hystorical Consequences. *IV Symposium Scienctifique International,* Paris, France, 25-26 Avril 2002, www.shroud. com/pdfs/aldo3.pdf

[38] Heller, J. H. and Adler, A. D. (1980). Blood on the Shroud of Turin. *Applied Optics, 19*, 16, August 15, 2742-2744.

[39] Jackson, J. P., Jumper, E. J. and Ercoline, W. R. (1984). Correlation of image intensity on the Turin Shroud with the 3-D structure of a human body shape. *Applied Optics, 23*, 14, 2244-2270.

[40] Jackson, J. P. (1990). Is the image on the Shroud due to a process heretofore unknown to modern science? *Shroud Spectrum International, 34*, 3-29.

[41] Jackson, J. P. (1998). *Does the Shroud of Turin show us the Resurrection?* Biblia y Fe.

[42] Judica Cordiglia, G. B. (1974). *L'Uomo della Sindone è il Gesù dei Vangeli?* Rome, Italy: Fond. Pelizza, Emmekappa.

[43] Judica Cordiglia, G. B. (1988). Ricerche ed immagini di laboratorio sulle fotografie eseguite nel 1969. In *"La Sindone, indagini scientifiche", atti del IV Congresso nazionale di Studi sulla Sindone,* Siracusa, ottobre 1987. Torino, Italy: ed. Paoline.

[44] Jumper, E. and Mottern R. (1980). Scientific Investigation of the Shroud of Turin. *Applied Optics,19*, 12, 15 June, 1909-1012.

[45] Jumper, E. J., Adler, A. D., Pellicori, S. F., Heller, J. H. and Druzik, J. R. (1984). A Comprehensive Examination of the Various Stains and Images on the Shroud of Turin. *ACS Advances in Chemistry, 205*, 447-476.

[46] Lattarulo, F. (1998). L'immagine sindonica spiegata attraverso un processo sismoelettrico. *III Congresso internazionale di studi sulla Sindone,* Torino, 5-7 Giugno.

[47] Latendresse, M. (2005). *The Turin Shroud Was Not Flattened Before the Images Formed and no Major Image Distortions Necessarily Occur from a*

Real Body. The Third Dallas International Conference on the Shroud of Turin, Dallas, Texas, September 8-11, 2005, latendre@iro.umontreal.ca

[48] Legrand, A. (1998). *Évangile et Linceul.* François-Xavier De Gilbert, Paris, ISBN 2868395074.

[49] McCrone, W. C. (1997). *Judgement Day for the Turin Shroud.* Microscope, Chicago, 287-288.

[50] Marion, A. (1998). Discovery of Inscriptions on the Shroud of Turin, Digital Image Processing. *Optical Engineering,. 37,* 2308-2313.

[51] Miller, V. D. and Pellicori, S. F. (1981). Ultraviolet fluorescence photography of the Shroud of Turin. *Journal of Biological Photography, 49,* 71-85.

[52] Miler, V. D. (1982). Quantitative Photography of the Shroud of Turin. *Proc. of Int. Conference on Cybernetics and Society IEEE,* October, 1982, 548-553.

[53] Moran, K. and Fanti, G. (2002). Does the Shroud body image show any physical evidence of Resurrection? *IV Symposium Scientifique International sur le Linceul de Turin,* Paris, 25-26 April 2002, http://space.tin.it/scienza/bachm/MORAN2.PDF

[54] Moroni, M. (1988). Ulteriore prova della presenza sull'occhio destrodell'Uomo della Sindone di una rara moneta emessa da Ponzio Pilato. In *"La Sindone, indagini scientifiche",* atti del IV Congresso nazionale di Studi sulla Sindone, Siracusa, ottobre 1987., Torino, Italy: ed. Paoline.

[55] Morris, R. A., Schwalbe, L. A. and London, J. R. (1980). X-Ray Fluorescence Investigation of the Shroud of Turin. *X-Ray Spectrometry, 9,* 40-47.

[56] Mottern, R.-W., London, R. J. and Morris, R. A. (1979). Radiographic Examination of the Shroud of Turin a Preliminary Report. *Materials Evaluation, 38,* 39-44.

[57] Nickell, J. (1987). *Inquest on the Shroud of Turin.* New Updated Ed.

[58] Pellicori, S. F. (1980). Spectral properties of the Shroud of Turin. *Applied Optics, 19,* 1913-1920.

[59] Pellicori, S. F. and Evans, M. S. (1981). The Shroud of Turin Through the Microscope. *Archaeology, January/February,* 35-43.

[60] Pesce Delfino, V. (2000). *E l'uomo creò la sindone.* Bari, Italy: Ed. Dedalo.

[61] Pia, S. www.nadir.it/pandora/Secondo-Pia/default.html

[62] Propp, K. E., Jackson, J. P. (1997). Color and Intensity Analyses of the Shroud of Turin. *Actes du III Symposium Scienctifique International du CIELT,* 45-50. Nice France.

[63] Rogers, R. (2002). *Scientific method applied to the Shroud of Turin*, a review. http://shroud.com/pdfs/rogers2.pdf, 2002

[64] Rogers, R. (2005). Studies on the radiocarbon sample from the Shroud of Turin. *Thermochimica Acta, 425*, 1-2 , 20 Jan. 2005, 189-194.

[65] Scheuermann, O. (1983). *Hypothesis: Electron emission or absorption as the mechanism that created the image on the Shroud of Turin – Proof by experiment* (first edition). September 1983, Milano, Italy: Fondazione 3M, Segrate.

[66] Schwalbe, L. A. and Rogers, R. N. (1982). Physics and chemistry of the Shroud of Turin, a summary of the 1978 investigations. *Analytica Chimica Acta, 135*, 3-49.

[67] Vignon, P. (1902). *Le Linceul du Christ.* Paris : Masson et C. Editeurs.

[68] Volckringer, J. (1991). *The Holy Shroud: Science confronts the imprints.* Manly, Australia: The Runciman Press,.

[69] Walsh, J. (1963). *The Shroud.* New York: Random House.

mapping, 207
Marx, 60
mass spectrometry, 148
materials, 9, 12, 13, 16, 40, 67, 71, 207, 211
matrix, 72, 89, 100, 120, 122, 125, 130
mechanical stress, 115
media, 42, 58, 61
medical, 54
medicine, 103, 154
medulla, 171
mercury, 164
meter, 6, 84
methanol, 33
methodology, 83
Mexico, 215
micrometer, 117, 151
microorganisms, 35
microscope, 38, 41, 127, 160, 170, 181
microspheres, 37
mitochondria, 38
mixing, 66
models, 207
modern science, 215
modules, 63
molecules, vii, 20, 39, 53, 54, 60, 67, 205
multiples, 65
mutation, 79, 82
myocardial infarction, 37
myoglobin, 37, 58

N

NADH, 30
nanometers, 151
national security, 35
negativity, 157
networking, 136
neural networks, 78
next generation, 131
numerical computations, 79
nuns, 144, 146, 147

O

OFS, 17, 18
oligomers, 59
opacity, 162
optical communications, 92, 102, 103, 110, 139
optical fiber, vii, viii, 7, 19, 20, 22, 37, 38, 41, 42, 56, 57, 59, 61, 62, 64, 65, 66, 69, 88, 93, 102, 103, 106, 123
optical gain, 68
optical properties, 13, 66, 68, 102
optimization, 62, 73, 77, 79, 82, 85, 93, 95, 102, 139
optimization method, 79
oral cavity, 166
orthogonality, 23, 24
overlap, 77, 104, 201
overlay, 181

P

painters, 153
paints, 181
palladium, 56
parallel, 14, 23, 124, 203, 204
pathogens, 34, 54, 60
pathology, 154
penalties, 86
peptide, 32, 37, 44
periodicity, 101, 109
permeability, 10, 13
pH, 46, 47, 48, 51
pharmaceuticals, 181
phase shifts, 6
Philadelphia, 19, 41
phonons, 67, 77
phosphate, 35
phosphorescence, 151
photoelastic effect, 107, 133
photographers, 179
photographs, 111, 127, 146, 150, 154, 155, 156, 157, 158, 159, 160, 161, 168, 176,